GROWING UP WITH

SPACEFLIGHT:

PROJECT GEMINI

Wes Oleszewski

Dedicated to Jules Bergman, who took us through
the ABC's of spaceflight by way of our televisions.

Series edited by
Jim Banke and Pat McCarthy
Technical Advisor Brian Nicklas

Some people got to experience America's space program up-close and personal, hands-on, steeped in the excitement of the peaceful advancement of human civilization.

The rest of us had to watch it on TV.

"Everything that we do in our adult life stems from something that we went through in childhood that other people blew off."

- My Mom, November 22nd, 2014

GROWING UP WITH
SPACEFLIGHT:
PROJECT GEMINI

GEMINI III
THIS WAS WAY BETTER THAN DICK AND JANE

It was a bitter cold Tuesday morning in mid-Michigan as we were all herded from Miss Schroder's second grade classroom and into the hallway of Nelle Haley Elementary School. Fortunately, energy was cheap in 1965 and the school's heat was cranked up. That was good for me, a scrawny seven-year-old asthmatic, because I needed the warmth to thaw me out after the agonizing walk from my home to the school's front doors; a distance of 1,183 feet. Of course when it was 19 degrees outside, that was uphill— both ways.

Following the teacher, I plopped my skinny butt down on the floor where she indicated and crossed my legs like a good little second-grader. Right next to me my pal Roger also planted his bottom. Roger and I had become friends the way little boys normally did in the mid 1900s— with a punch in the mouth.

At that time, it was the first week of school and I was the new kid in Nelle Haley as my family had just moved into our Sheridan Park subdivision in the mid-summer of 1964. Roger figured to pick on the new kid during recess one day and began by calling me "Wesbee" rather than Wesley. One thing led to another and I took a swing at Roger. He retaliated by letting me have a knuckle sandwich that only lightly connected.

What he didn't know was that I had a loose front tooth that I had been wiggling for weeks and his punch had just finished the job. A bit stunned I stepped back a little with my dukes still up. I felt the tooth in my mouth rattling around and I could taste the blood. Not being much of a fighter, I was pretty quick witted and knew instantly what to do. I took a step forward, spit the tooth on him and then stepped back and smiled showing the gap. He looked at me, looked at the bloody tooth now stuck to his shirt and ran off to the teacher screaming! All of the other kids laughed, but Miss Schroder was not amused. She dragged me into the classroom, gave me the tooth wrapped in a tissue, a good scolding and told me that when I went home I was to tell my parents what I had done.

Oddly, on the walk home from school that afternoon, Roger and I became friends. Since he thought it was all pretty funny, I figured there was no reason to bother Mom with what happened and I did get a nickel for the tooth after putting it under my pillow. By the time of parent-teacher conferences Miss Schroder had apparently forgotten all about the tooth incident, because my Mom came home and never mentioned it.

Now, seven months after Roger helped me lose that tooth, we were sitting together trying to figure out why everyone in the school was out in the hallway.

From a nearby closet Mr. Leady, one of the teachers, had one of the older kids in his 6th grade class roll out a TV set atop a big brown metal stand. As he plugged it into the wall outlet I was in hope of seeing cartoons, but morning cartoons were over by this time in the morning.

Turning the set on he messed with the "rabbit ears" antenna and in short order a black and white image popped onto the screen. We all squinted as he stepped around and looked approvingly at the TV.

"What is that?" I whispered toward Roger.

"I dunno," he shrugged.

Suddenly one of the older kids spoke up and informed the murmuring crowd that there was a space launch today and we were going to watch it!

"A space launch!" Roger gasped at me.

"Like Fireball XL5!" I exclaimed back excitedly.

This was great! Nothing better to save us from being forced to read those non-thrilling adventures of Dick and Jane and their amazingly boring dog Spot than a space launch. "See Spot run. Run, Spot, run." Talk about child abuse; that was it.

Of course, there were some problems, the rocket on the TV looked nothing like Fireball XL5. Plus it was standing upright and Fireball XL5 launched on a horizontal track. The rocket on the TV was skinny and stood next to a tower. How in the heck could that thing get into space? Beside all of that, Fireball XL5 was on TV on Saturday and not weekdays.

Mr. Leady told us that this one was called Gemini 3 and it was going to send two Americans into space. I wondered if they went aboard on hover cycles like the guys on Fireball XL5.

(MEMORIES OF OTHER FOLKS) Dave; Saginaw, Michigan: "We watched the Mercury program at Nelle Haley. Whenever one went up, we saw it live. When I had Mr. Leady as a teacher, he appointed me the audio visual kid. I had to get out the film strip projector and thread those boring slides into it and then make it advance one with the recorded voice-over and then I graduated to actual movie work, and then he had me picking out the movies we were gonna watch... you know...the history and nature films... I would order them and the school district would magically make them appear." Dave was probably the guy who rolled out the TV so we could all watch the Gemini 3 launch.

All of the adults were quite excited about this "space launch" and so were the people on the TV set. The fact was this was the first manned launch by the United States in nearly two years and most adults were well aware that we were in a space race with the Soviets. Now a whole bunch of American spaceflight firsts were going to take place in a matter of just a few hours. A large portion of those firsts would occur just with the launch itself.

For the first time we would be launching a two-man spacecraft. It was brand new and had never flown with a crew aboard prior to today. It was stacked atop a new booster for manned flight- the Titan II. Nearly every system in the spacecraft was different from those in Mercury or was a greatly modified Mercury system. This was the first US crew to fly atop a two stage booster and if it blew up their lives depended on ejection seats rather than an escape tower. Finally, this was the first manned flight to liftoff from Launch Complex 19 at Cape Canaveral. So, of course the adults were worried, but I was more concerned with whether or not I had left my milk money at home that morning. A quick check of my pockets showed that the nickel was there— I could relax.

The super-imposed countdown clock on the TV

screen was steadily counting backward and I had no idea what that meant. What I did know for certain was that sitting out in that hallway was far better than reading "Dick and Jane," especially when there was a TV set involved. The reality was that America was within minutes of hurtling a significant milestone in our space program and all around the nation countless kids like me were seated in schools and watching it on TV. It would be a memory that I would always carry with me— unlike my milk money.

GEMINI III
MOLLY BROWN'S BRIEF
INTRODUCTION TO THE NATION

Second grade for me was a pretty simple adventure because the mid 1960s were a fairly simple time.

Our subdivision of Sheridan Park was a rapidly growing community on the industrial side of the city of Saginaw, Michigan and most of the dads had industrial jobs.

My Dad was a railroad engineer on the C&O and for me going to the "yard" and climbing aboard a locomotive was simply going to where Dad worked. All around us the factories hummed 24 hours a day, 365 days a year. We lived from paycheck to paycheck while the cupboard grew thinner of groceries as the 14 days until the next payday counted down. Milk was delivered to our door in half gallon glass bottles filled at a local dairy, bread was baked at a local bread making plant and our second grade class did field trips to both. Such field trips were a fun adventure considering that the dark winter consumed most of the school year and it was a real treat to extend your world beyond the school room and your house.

Often we were snowed in to the point where just walking between home and school was a chore. "Snow days" were very rare and reserved for monumental blizzards. Of course energy was cheap in those days and both school and home were always warm. "Cabin fever" was avoided by Mom kicking us outside into the snow to play as soon as we got home from school. You could come back in when you could no longer feel your hands and feet.

Our 1,014 square foot house at 3324 Lexington Drive was a single story three bedroom home with one bath, no shower, a fair sized living room, small dining area and nice sized kitchen. It was standard size for Sheridan Park. We were warm, happy and well fed... up until a couple of days before payday.

NASA and Project Gemini were pretty far away from industrial mid-Michigan; right up until 10pm Sheridan Park time on Monday, March 1st, 1965. That was when the two most trusted men in America, Walter Cronkite and John Glenn, came onto many of our neighborhood TV sets by way of CBS News.

My Mom allowed me to stay up that night because I had just put them through a fairly shocking experience a few months earlier. I had just started second grade in the fall of 1964 when my parents were called in to meet with my teacher and were given some stunning news...

"Wes can't read."

My astonished parents were informed by my teacher that I was unable to read aloud those captivating adventures of Dick and Jane and their fascinating dog Spot. They were told that they had to take me downtown to the Board of Education building to meet with a "Reading Specialist." In 1964, such a visit to such a person was a real embarrassment- something to be done quietly and in the shadows— because it meant your kid was somehow defective. If the word got out, the Moms would gossip and the other kids would tease.

Thus, my parents meekly took their defective son downtown. We entered the big, dark, old brick building with its echoing hallways and met the lady who was "The Reading Specialist." She had been waiting for us and was very nice as she led us to a doorway that opened to a big room. She directed me to go inside and wait while she had a few words with my parents. The room was a former high school classroom that now had living room-styled furniture in it and was filled, (and I do mean FILLED,) with books. The books were seemingly tossed around everyplace and were about all sorts of subjects. There was also one, almost unnoticeable feature in that room- a very large mirror on one wall.

I walked around the big room for about 90 seconds and I was bored. Thus, I found a book on the Boeing 707. I sat down with it and started reading. Unknown to me, behind the mirror at the end of the room was the amazingly savvy "reading specialist" and my worried parents— watching my every move.

"See..." she told my folks, "he can read— he's just bored with Dick and Jane."

She also informed them that the book their scrawny second-grader was currently reading was on the sixth grade reading level. My Mom asked about my teacher's concern that I "could not read aloud." In the following discussion my folks were informed of the difference between "Can't" read and "Won't" read. Interestingly, I'm currently the author of more than 20 books and I still refuse to read aloud to the public… ever.

Oddly, more than a half century later I have one surviving aunt who still thinks that I am totally illiterate. Yes, she is convinced that I faked my way through high school and two years of advanced electronics, a four-year degree in Aeronautical Science at the Embry-Riddle Aeronautical University (the toughest aviation school on the planet), a profession as a flight instructor, airline Captain, corporate jet pilot and author. Some stigmas are hard to shed when it comes to certain people.

My parents were given marching orders to get me anything, ANYTHING to read in which I was interested. Comic books, Mad Magazine, science books, sports books— anything that I was interested in should be provided. Likewise, television shows that were of an informative nature were to be allowed along with the encouragement that more can be found out about the subject through reading. Mom and especially Dad were then spring-loaded to respond to what sparked an interest in me, and the CBS News program hosted by Cronkite and Glenn was just the ticket.

In Saginaw that night CBS was broadcast on UHF Channel 57, WKNX. Both the station and the main antenna were within 1,180 feet of my living room TV set. Although 10:00 in the evening was a bit later than I normally stayed up on a school night my parents were quite sensitive to anything that was informative or remotely educational either in print or on TV that may have the slightest chance of holding my fleeting attention. Thus, I got to stay up "late."

CBS News had a weekly slot open in their 1965 schedule on Monday evenings at 10:00 where they would provide "various" programming. On March 1st of that year the program presented was titled "T Minus Four Years, Nine Months and 30 Days." Although the program was primarily about Project Apollo, the show did a segment on the upcoming Gemini mission. They were up in the white room at Launch Complex 19 where Cronkite introduced "Man's next step in space" the Titan II booster and atop it- Gemini 3.

The Gemini segment ran four minutes and 23 seconds and started off with John Glenn's former test director John Williams in the white room. Next to him was MOLLY BROWN, the spacecraft that Gus Grissom and John Young would fly just 22 days later. The hatches of the spacecraft were wide open and although the telecast was in black and white you could almost see the color of the "Remove Before Flight" tags hung upon it. After Glenn and Williams finished some long distance, "How ya' doin' " chatter, Cronkite asked Mr. Williams "What is the Gemini Mission? What are you going to find out on the Gemini flights?" In his answer Mr. Williams explained that the purpose of Gemini was to gain knowledge on extended weightlessness and its effect on the astronauts, rendezvous and dock with unmanned target vehicles and develop the capability of active reentry control. Of course to a seven year old kid that all went past like ice flecks off of a spacecraft. All we wanted to see was the Gemini and its insides.

Soon the camera panned around and there were two "crewmen" seated in MOLLY BROWN doing some inertial guidance system checks. Neither of the supposed crewmen looked like Schirra or Stafford, the back-up crew for GT-3, and it is a good bet that Grissom and Young were far too busy to be doing that sort of dog work that close to the launch date. Compared to the cramped little Mercury capsule, the Gemini spacecraft looked huge and those open hatches looked like barn doors! (See Image 1, page 157.)

Next the show switched to inside the LC-19 blockhouse and US Air Force Lt. Colonel John Albert, who was in charge of the USAF end of launching the Titan II booster. This was a space-buff's first look inside a working blockhouse at Cape Canaveral. He spoke about the guidance systems, but behind him were TV monitors showing all sorts of different views of the GT-3 vehicle plus lots of technicians with headsets on and shirts that had "Martin" or "McDonnell" embroidered on the back— it was all so cool, even in black and white. Certainly, any 7-year-old would have been glued to the TV that evening. That is especially true considering that all the other two networks had on in that time slot was "Ben Casey" and "The Alfred Hitchcock Hour."

That Cronkite program made TV history in another way as well. For the first time every segment was filmed on the same day that the show aired. Segments were recorded from Washington DC, Downey, California, Houston, Texas, Huntsville, Alabama and Cape Canaveral. The video tapes were then sent to New York and quickly edited and shown that same evening on this program. It was an experiment that worked out quite well as the program appeared to be almost done live.

When the show was over my Mom asked what I thought of it. I gave some sort of positive answer knowing full well that unless I reacted positively I may not get to stay up late again and watch TV. Of course in my Swiss cheese attention span I did not really connect that program with the vehicle that we were looking at when we sat out in the hallway waiting for Gemini 3 to launch.

GEMINI III
THE S-3 EXPERIMENT

To command the crew of Gemini 3 the chief of NASA's astronaut office, Deke Slayton, had selected his best astronaut- Gus Grissom. "A hard workin', hard playin' straight shooter" is how Grissom's teammate for the flight, John Young, would later describe Grissom.

Originally the crew of GT-3 had been Alan Shepard and Tom Stafford, but an inner ear malady had knocked Shepard off flight status. Grissom and Frank Borman had originally been assigned to GT-5, but as soon as they were paired for the mission it became obvious that they did not get along well. Some had said, tongue-in-cheek, "they'd probably kill each other." So, Deke moved Young into GT-3 as "Pilot." (Among the astronaut corps everyone thought they were the best pilot, so no one wanted to be called a "co-pilot." Thus, the proper titles for the crew positions on Project Gemini became "Commander" and "Pilot.") The two astronauts were the perfect combination to do the flight test of the first Gemini spacecraft.

Everyone knew that the spacecraft would certainly have a few bugs in it and Grissom and Young, being good test pilots, were set to wring out every single bug. But they were not expecting sea urchins.

At T-minus 20 minutes and counting, just as my

second grade class was being herded out into the hallway to watch the launch, the crew of GT-3 were conducting a test firing of the reentry Reentry Control System (RCS). Located on the snout of the Gemini spacecraft the reentry RCS jets were arranged in two rings each having eight thrusters. Each ring was an individual system thus allowing redundancy.

Young did the switchology using the toggles just to the left of his left knee, while Grissom systematically burped each thruster with the hand controller located on the center pedestal in two bursts lasting about one half of a second. Only the "B" ring was fired for the static test and then it was turned off and left for flight. The "A" ring was not fired on the ground. That way the "B" ring could be used in the case of a high altitude abort.

Following that test the Orbital Attitude Maneuvering System (OAMS) was also fired in a similar manner. Each firing consisted of the Spacecraft Test Conductor (STC) in the blockhouse reading a checklist, one item at a time, and then the crew performing that task and confirming it to the STC who double checked with his Propulsion officer. It was a very methodical and careful process that has always been the backbone of US human spaceflight.

One thing astronauts loathe is last minute changes. Crews train for years for a specific mission so that all tasks can be done correctly and last minute changes negate that effort. One thing scientists love is last minute changes- because it means that they can get their egghead experiment onboard a flight.

After nearly two years without a US manned spaceflight, the scientists just could not let GT-3 be the pure flight test that it was supposed to be. So, on GT-3 the last minute changes were many and the final one was the "S-3 experiment." The crew caught a technician installing a new black box in MOLLY BROWN and when Gus asked what it was he was told that it was set up to test the synergistic effect of zero gravity on sea urchin eggs.

Gus was not happy.

He was informed that all he had to do was turn a handle at the appointed time to fertilize the eggs. Of course that action alone likely prompted some imaginative teasing among not only the astronauts, but the ground crews as well. Just past the T-12 minute mark in the countdown the following exchange took place over the net:

"CM-1, STC." George Page called out over the net.

"Go ahead." Gus answered.

"Gus can ya' tell whether or not those sea urchins are up to anything yet?"

"They're thrashin' around back there, must be doin' some good."

"Great," STC replied and as he paused a protracted period of low laughter is heard over the net.

"You don't suppose they hatch like rabbits do ya'?" Gus retorted in his normal deadpan style.

"Gee I hope not. It'll be crowded in there." STC commented with a hint of snicker.

"Yeah." Gus quipped.

As it would turn out, even the S-3 sea urchin experiment, where all you had to do was turn the handle, would have its bugs.

GEMINI III
YEAH MAN

As the countdown clock on the school's TV set hit the T-10 second mark the teachers began reflexively "shooshing" us. By the time our chatter quieted only a couple of seconds remained in the countdown. We heard the voice of NASA's Paul Haney as he counted off the final two seconds down to zero,

"Ignition," he announced confidently.

On our black and white TV set a billow of white steam rapidly grew behind the Titan II launch vehicle, then it suddenly began to move.

"And we have liftoff," Haney stated with a touch of glee, "...we have liftoff at 24 minutes after the hour."

Oddly, there appeared to be very little flame coming from the rocket. All we saw on the TV was a short, translucent spike flickering from the engine bells.

Unlike the Mercury Atlas and even Fireball XL5, the Titan II burned a hypergolic fuel and oxidizer. A 50/50 combination of unsymmetrical dimethyl hydrazine and hydrazine, later renamed "Aerozine 50" by Aerojet General, was the fuel and a mixture of nitrogen tetroxide and nitric oxide was the oxidizer. Both the Aerozine 50 (also seen listed by some as Aerozene 50) and the oxidizer would burn spontaneously on contact. When they did burn they made a nearly clear flame and no smoke. These propellants, however, were extremely toxic— which meant little in the space race of the 1960s.

Upon liftoff both astronauts could hardly feel the Titan II boosting beneath them. The vehicle was much smoother than expected and there was far less noise than had been presented in the simulator. Just 26 seconds after liftoff CAPCOM Gordon Cooper, one of the original Mercury astronauts, called out to Grissom, "Yer' on yer' way Molly Brown,"

"Yeah man." Grissom replied as if sitting in an easy-chair in his living room.

Of course the public heard none of that directly because all conversation between the astronauts and Mission Control went no farther than Paul Haney's headset. He then repeated what was being said and filtered out anything that NASA may not want heard. It was a process that would not change until Apollo 10 when live conversation between the crews and the ground would finally be permitted.

"Molly Brown" referred to the name the crew had given their spacecraft. In his book "Gemini", Grissom states that he had originally wanted to christen his spacecraft with an "Indian name" so it would be "truly American." Asking the researchers at "Life" magazine and "World Book Encyclopedia" to find such names he chose "Wapasha" which was the name of a tribe in his home state of Indiana who used to camp on the Wabash River. Then it was pointed out to him that some joker would sooner, or later re-dub the spacecraft "The Wabash Cannon Ball." The name comes from a song about a mythical train from 1882. The train that ran for a mythical railroad was said to be everything from a death train to one that went into space. Grissom's Dad worked for the B&O railroad and Gus had second thoughts about what his Dad would think if the spacecraft was not named after a B&O train. Additionally, NASA would not be too happy about the cannon ball aspect of it. Thus, Grissom dropped that name and went looking for another.

It occurred to Grissom that there was a popular musical that was just finishing its run, "The Unsinkable Molly Brown." Gus selected his name to spoof the fact that his Mercury spacecraft had sunk to the bottom of the Atlantic Ocean after splashdown. The moniker itself was taken from a musical written about Margaret "Molly" Brown, an American socialite who had been a passenger aboard the TITANIC and survived the ordeal.

Some managers in NASA, however, were not at all happy with the idea. When one sent a letter to Grissom asking what his second choice would be, Gus replied with "TITANIC." Grissom and Young did, however, shorten the name to simply "MOLLY BROWN." NASA management felt a bit better, but put a stop to crews naming their spacecraft. That policy stuck until Apollo 9 when two spacecraft, the Command Service Module and Lunar Module would be flying separately in orbit at the same time. Then individual spacecraft names would be required for communication purposes.

While GT-3 climbed away from Cape Canaveral all I saw on the school's TV set was a pencil thin rocket that kept getting smaller and smaller until it was just a dot. Of course I was sitting about 50 feet from the TV so there was no need to squint, the whole show had turned into just news men and "the voice of NASA" Paul Haney chattering about stuff, the meaning of which I had not a clue. Yet it was still better than Dick and Jane. (See Image 2, page 157.)

(MEMORIES OF OTHER FOLKS) Tim: "I was glued to the TV for nearly every launch and event. I was 10 when Gemini 3 launched and remember many of the highlights as I watched on TV. I used to keep a box of newspaper clippings but they were accidentally thrown out during one of my parent's cleaning frenzies."

GEMINI III
URCHINS IN SPACE!

Some 50 seconds into GT-3's climb aloft the crew felt the speed building as well as the "G" loads. Now the spacecraft began vibrate and shake heavily as the whole stack pushed through the sound barrier and into the area of maximum aerodynamic pressure (Max-Q). MOLLY BROWN shook as if every rivet and bolt was being tested. An aerodynamic shock wave was forming along the vehicle. At 01:01 into the boost the Titan II had punched through the speed of sound and 16 seconds later Max-Q was passed. All of the vibration and shaking rapidly diminished. Suddenly the crew was wrapped in silence as the G-forces began to press harder and harder upon them. The sky outside their windows began to grow black as Gemini 3 left most of the Earth's atmosphere behind.

Now as the fuel was being burned off and the vehicle rapidly became lighter, the Titan II began to accelerate at a rapidly growing rate. With the rapidly building G-load, Young later recalled that he realized at that point that he was just along for the ride because, "…that baby was really tearing along."

(See Image 3, page 158.) In their seats Grissom and Young were pulling 6.5 "G's" by the time the Titan II's twin Aerojet LR87-AJ-5 first stage engines began to starve themselves of propellants. It took less than two seconds for the staging sequence to take place. Sensing the first stage engines' chamber pressure dropping below 65 percent, arm relays were activated. The staging control relays were next activated and in turn the staging switches were closed by the system. Then the stage I solenoids were energized and both of the first stage engines were muted. For an instant the G-load dropped off to 1.5 Gs and the two astronauts were shoved forward against their straps. Next the booster sequential system commanded ignition of the second stage's single LR91-AJ-5 engine while the two stages were still connected to one another.

Around the inner ring of the separation plane a series of explosive bolts holding the first and second stages together detonated. The force of the exhaust from the second stage engine then literally blew the two stages apart as most of the exhaust was vented through openings in the interstage, yet the upper section of the interstage was constructed of blow-out panes that simply flew away as the two stages parted. For the first time astronauts witnessed the flash of the Titan II's staging fireball as it flashed above their windows and then the booster propelled them right through it.

Once the crew was riding on the second stage the active guidance was initiated and the vehicle began seeking its assigned course. As the crew watched, the nose of the spacecraft dropped below the horizon and then came right back up again. "We're starting to steer," Grissom reported calmly to the ground, "The horizon comes right into view."

One of his duties as a test pilot at this point in the

mission was to evaluate the Commander's visibility through the Gemini spacecraft's window. In fact, in preparation for Project Gemini, that window shape had to be tested— in flight. So, it was Gus who was assigned to fly a Piper Apache light twin propeller aircraft with all of the windows blacked out except for the front, which had a Gemini window-shaped area left clear. Gus flew the aircraft and landed it reporting, "No problems whatsoever." Years later he got to evaluate the window again- this time, in space.

Acceleration on the second stage was far greater than the first stage and in short order both Grissom and young were soon pulling 6 Gs. For another three-and-one-half minutes the second stage raced toward orbit until cutoff.

The astronauts were now supposed to be up to 6.8 Gs, but the GT-3 booster slightly over accelerated to 7.9 Gs and then suddenly went to zero-G at cutoff!

Grissom watched his event timer and thrusted the spacecraft for 20 seconds after cutoff then Young pushed the SEP SPCFT switch in the upper left corner of the main panel. That fired the detonators intended to separate the spacecraft from the second stage. The action set off two flexible linear shaped charges that ran in a circle around the mating ring that held the spacecraft to the second stage. That explosion blew apart the mating ring and effectively severed the spacecraft adapter from Titan II leaving behind the lower portion of the mating ring that had bolted the spacecraft to the booster. Their detonations sounded like a round of cannon fire in the cockpit and once free, Gus thrusted MOLLY BROWN forward for a bit more than 12 seconds.

Young saw white material from the separation, "like ping-pong balls" shooting by and fanning out in front of the spacecraft. America's first two-man crew was in orbit.

At Nelle Haley Elementary school we were being herded back into our classroom like so many sheep. While the rest of grown-up America were thrilled to watch the coverage of Gemini 3, we were being led back to those thrilling adventures of Dick and Jane… and of course, Spot the dog.

Just before Gus' event timer hit 00:20:00 into the mission and the crew was busy running their checklist, MOLLY BROWN had taken on a tendency to slowly yaw left. Gus began thinking he had a thruster leaking.

Suddenly the Canary Island ground station's CAPCOM called up with a "Mark!" for the 20-minute point. It was a cue for Gus to fertilize his sea urchin eggs. Dutifully, the commander reached up and turned the handle to do the dirty deed. Although Gus thought that he had successfully fertilized the eggs, the handle did not seem to turn in his hand as it should. With his spacesuit's gloved hand it was impossible to tell how much tension was actually being applied to the handle of the experiment.

In fact the people who designed the experiment had given little thought to turning that handle in zero gravity and unknown to Gus he had not successfully fertilized the eggs, but had actually broken the handle's internal fitting. He was supposed to turn the handle two more times during the flight and did so on schedule, but, again unknown to him the exercise was pointless.

The experiment was coordinated with a scientist on

the ground, who would perform the exact same action as Gus at the exact same moment. That way when the mission was over the biologists could look at the two sets of hatched sea urchins and, comparing those fertilized in space to those fertilized on Earth, see what their differences may be. Oddly, when the scientist on the ground turned his handle, it also broke off in his hand. The entire experiment was a failure due to the poor design of the package.

GEMINI III
LOOK AT THE CABIN PRESSURE!

Just four minutes and four seconds after Grissom had not fertilized the sea urchin eggs, Young startled the commander when the normally stoic pilot exclaimed,

"Look at the cabin pressure!"

Glancing down, Gus saw that both the cabin pressure and the suit pressure were reading zero!

"What?!" Gus replied

"We lost the…" Young began.

Instinctively, Grissom reached up and slapped shut his helmet visor.

"Lost the what?" Gus asked in a commander's fashion, "What've we lost?"

Then it struck him that those pressures could not actually be zero, or both he and Young would already be dead. And even if the reading was correct, closing his visor would not make any difference. Sheepishly, Grissom reopened his visor as Young quickly did some mental trouble-shooting and determined that the primary electrical DC-DC converter had failed.

"Lost the primary converter," Young replied as he switched to the back-up converter and the readings came back to normal.

It was a moment when the old dog hardnosed test

pilot Grissom got a good taste of John Young's super cool in space.

Of course the crew still had that persistent left yawing of the spacecraft to deal with. Mission Control thought they had a leak in a thruster and Grissom thought they had a leak too.

The crew, however was far less concerned about the slight yawing then were the ground controllers. The fact that communications with the ground by way of remote stations was often ratty did not help the situation. Circuit breakers were cycled and thruster logic was altered and the yaw continued. At times the crew thought that the tendency had ceased, but then it would creep back in. It would eventually end up as a simple footnote in the historic mission. A minor annoyance to the two astronauts, yet a big sweat to Mission Control.

At 01:33:00 into the flight, GT-3 made spaceflight history in a big way. It was at that moment MOLLY BROWN's Orbital Maneuvering System (OMS) thrusters fired up in retrograde and burned for 74 seconds. Gus saw a bolt come out of hiding behind him and pin itself to the instrument panel then float away when the OMS thrusters cut off. The result was a change in the orbit of GT-3 as it was lowered by 34 miles.

No manned spacecraft had ever actively changed its orbit prior to Gemini 3. The Mercury capsules could not do it and the Soviet Vostok could not do it as neither had OMS thrusters. The Soviet Voskhod, which was little more than a Vostok with additional seats and no escape capability and was flying as the rebuttal to Gemini, also had no orbital change capability.

So, this burn plus another one made at 04:14:00 that shifted the spacecraft's orbital plane were what should have been the most remembered part of the GT-3 mission. But they were about the be overshadowed by a sandwich.

GEMINI III
SMELLS, DOESN'T IT?

Beginning at 01:40:00 into the mission, just over four minutes into the second orbit of GT-3, the timeline called for Young to begin acting as ship's cook. He was to unstow items of food and test not only the taste and handling of the food, but also the packaging.

NASA had gone to great lengths to try and create food items that would function in zero gravity. Mercury astronauts were the first orbiting Americans to try and eat in space and actually managed to do so although the "food" itself left a lot to be desired. Cubed items such as brownies and bacon worked in space as did assorted food mutations that had been made into a format that could be squeezed from a tube.

Originally, many in the medical community thought it would be impossible for astronauts to swallow without gravity assisting the process. That myth soon went away along with many other medical misconceptions.

By Gemini 3 the art of space food had developed somewhat from the Mercury days. Grissom and Young's meal consisted of chicken "bites," reconstituted apple sauce and grapefruit juice. But Young had an extra item in mind. Reaching into one of the pockets of his spacesuit he pulled out a genuine Kosher corned beef sandwich! In the transcripts there is no record of Young actually saying anything to his commander when handing him the sandwich.

"What is it?" a puzzled Grissom asked.

"Corn beef sandwich," Young replied calmly.

"Where did it come from?" Grissom questioned, taking the snack.

"I brought it with me," Young explained in his deadpan manner, "Let's see how it tastes. Smells, doesn't it?"

Grissom took a couple of bites, but saw that crumbs began floating around and decided to stow the sandwich.

"Yes," the commander noted, "it's breakin' up. I'm gonna stick in my pocket."

"Is it?" Young responded, "It was a thought, anyway."

"Yep." Grissom quipped.

"Not a very good one," Young half apologized.

"Pretty good," Grissom supported his rookie partner, "though, if it would just hold together." He stashed the contraband sandwich in the left leg pocket of his spacesuit.

In fact, Grissom later stated that Young's offer of the "strictly non-regulation goodie" always remained one of the highlights of that flight for him.

According to John Young, the fact is the sandwich, corned beef on rye, no pickle and no mustard, was made for Wally Schirra at Wolfie's deli located on North Atlantic Avenue in Cocoa Beach. Schirra, the back-up commander for GT-3, brought the sandwich with him to Launch Complex 16 where the GT-3 crew were scheduled to suit up and discreetly handed it to Young who stashed it in the pocket of his spacesuit.

After the mission the story was leaked to the press and a firestorm of controversy that went all the way up to a Congressional hearing erupted. It was implied that the crew had ignored their testing of proper space food in favor of a corned beef sandwich. Proper space food that NASA had spent an estimated $30 million developing. Of course it was all political grand-standing primarily on the part of Congressman George Shipley who called it the $30 million sandwich.

The over-blown nonsense nearly overshadowed the entire Gemini 3 mission. That, by the way, was not Schirra's only prank for the GT-3 mission. When Grissom and Young arrived for their suit-up, Schirra met them wearing an old torn-up and tattered Mercury pressure suit. He shook Gus' hand and said that he was ready to go in case they decided to back out of the mission. Fortunately, Congressman Shipley did not get wind of that one.

(See Image 4, page 158.) In the early 1970s my family took our first Florida vacation and we stayed at the Atlantis Motel that was near the Ramada Inn where Wolfie's deli was located. Of course I dragged my Mom there for lunch. While we were selecting food, Mom was explaining to the guy behind the counter that we had just been to the space center and how I was such a huge space-buff. The guy looked down at my scrawny 15-year-old self and said,

"I suppose you'll want the corned beef sandwich?"

I was served a corned beef on rye and the deli guy explained that Gus' didn't have mustard, but I could find that on the table. I'm not a fan of rye bread, but I ate that one. Wolfie's no longer exists, the spot where the deli stood is now a gift shop in a large convention complex.

Of course Young dutifully went about testing the proper space food. Reportedly the chicken bites tasted like cardboard, but the apple sauce was pretty good and Grissom tried some of that.

"If we had some pork chops to go with it, we'd be alright," he commented.

A half century later, astronauts aboard the International Space Station maintain a continual human presence in Earth orbit— and they do have pork chops.

Young's duties then shifted to the other end of the bodily functions spectrum— an area known in spaceflight speak as "waste management." In commencing this part of the flight plan, John Young would achieve another American space first. Yet, this record in spaceflight history would not be one that he would want to boast about during his career.

Considering that future Gemini flights were scheduled to last between eight and 14 days NASA had to invest some serious time and money in the problem of going to the potty in space.

Considering also that my own children will be reading this book and in this age of so many people being spring-loaded to the "I'm offended" position I want to keep things readable for as many people as I can, this text will use the simple terms of "pee" and "poop" to further this subject. In fact, they actually require less typing that "urination" and "defecation." Thus, I need not explain them to my kids or anyone else.

The most asked question about spaceflight has always been, "how do they go to the bathroom up there?" In fact, prior to Skylab in 1973, there was no such thing as a working toilet in space. Additionally, prior to Gemini in 1965 there was no need for such facilities because American space flights were very short. Astronauts were fed a low residue diet for several days prior to a mission and nearly every crumb of what they were fed was completely digested— so there was no poop to deal with.

Mercury astronauts, after Alan Shepard's need to empty his bladder aboard FREEDOM 7, had to address the issue of "Where do you pee?"

Gus Grissom was given a sort of adult diaper attached to a garter belt to wear under his spacesuit to handle any pee aboard LIBERTY BELL 7. Later Mercury flights allowed the astronaut to pee into an in-suit storage bag- with the exception of Gordon Cooper's FAITH 7 mission. Cooper had a hose attached to condom-like device that allowed him to pump the pee into a storage container. It did not work well and Cooper simply drank less water and gritted his teeth until he was on the ground.

I am not going to cover how the Soviet cosmonauts handled the issue, because they did everything in secret, so who cares? Frankly, as far as we in the free world knew, their crews were probably just ordered to hold it— OR ELSE!

By the time the United States began Gemini, the problem of going potty in space had turned into an actual "problem" and on Gemini 3 it got worse.

NASA's solution to the issue of going pee involved something similar to Gordon Cooper's Mercury "receiver," only this one had an inflatable cuff to ensure a proper fit to the astronaut. The cuff was pumped up by an external hand pump and when the astronaut let go with the flow it was supposed to go into an extendable bellows that was in turn connected to an overboard dump valve. Along the way a urine-sample valve had been installed so some pee could be caught and returned to earth for doctors to study.

In order to use the device on Gemini 3, Young had to unzip his suit. He found that while doing that the suit's lower strap was in the way and he had to use the scissors that he had used earlier to open the food packages to cut the strap. Once he was properly positioned and tried to pee, the flow tended to not go into the bellows and some leaked back into the cabin and floated freely around before Young was able to re-cap the cuff.

Oddly, NASA got all worked up over crumbs from a corned beef sandwich when the real worry should have been about the droplets of pee. Urine is a great conductor of electricity and could likely do far more damage and get into much more confined spaces than any bread crumbs.

Next on the list for the waste management designated test objective was, of course, poop. Apparently Young had been spared the required diet of low residue foods because it was part of his mission to test the "Defecation Collection Device" later known as "The Blue Bag."

This was the most simple device ever conceived for spaceflight and it consisted of a large plastic bag with the open end officially described as having, "…a sticky surface used to seal the bag to the buttocks."

The astronaut needing to poop would get the proper part of his body out of his spacesuit and then position the bag properly and… poop. It was just that simple—okay, it actually wasn't.

Everyone here on Earth takes for granted the power of gravity and no matter if you are a hamster or a dinosaur, since the beginning of time gravity has done all of us a great favor; it allows poop to simply drop away from the body. No matter if it is into an outhouse, or a toilet, or simply into the woods, gravity takes the poop away from your body.

In space, however, gravity does not work that way-instead you and your poop keep orbiting… stuck together. The poop does not drop anywhere- it just piles up at the point of exit. Thus, John Young found himself with a substantial mess and not enough paper and wipes to really clean it up very well. Fortunately for Grissom, the scrubbers in the Environmental Control System removed nearly all of the odor. Unfortunately for Young, the entire waste management exercise took much longer than scheduled. Young even complained that he got some on his glove. It was the price to be paid for achieving a US space "first."

The In-Flight Urine Management System used on Gemini 3 was later scrapped, although the Gemini 4 mission was scheduled too soon to receive the new system and it flew with the old GT-3 system.

A new system with a cuff attached to a condom-like roll out fitting and an 800cc collection bag equipped with quick release fittings replaced the bellows. Gemini missions from GT-5 to GT-12 used the new system with great success.

All of the missions, however, were stuck with

the Blue Bag for pooping needs. Yet, for all of those troubles, John Young became the first American to poop in space. He may have even been the first man to poop in space, but we are not sure about the Soviets. Again, my guess is that they just held it as ordered… it's a Soviet thing.

GEMINI III
FLIGHT TEST NIT-PICKING

Above all, GT-3 was a test flight and although the crew later officially said that the mission was "…pretty uneventful…" they actually uncovered a lot of little snags and learned a lot about the Gemini spacecraft itself.

Most annoying problems that the crew uncovered were actually due to poor planning by NASA. Although the crew had been in training for nearly two years for this type of a mission, a lot of simple things had not been considered. The worst seemed to be that the crew had been given nothing to write on! There was no scratch paper or anything to take notes on. Grissom solved some of this by taking out his Pentel pencil and actually writing things on the knee of his spacesuit. After the flight both astronauts recommended that crews be given kneeboards on future missions.

Amazingly, neither Grissom, nor Young had been proficiently trained on the use of the 16mm camera that was flown aboard GT-3. Thus, when it came time to take pictures they had to guess what aperture setting were correct for a given lighting condition.

Also the crew found that the handy-dandy Flight Plan Roller that was located in the upper middle of the center instrument panel was nearly useless. The panel itself was poorly lit and the writing on the rolling strip of flight activities was so small that it could not be read. In fact, some notes that Grissom and Schirra had written on the strip itself could not be read. Additionally, looking at what was presented in the roller's window was like looking at the mission through a keyhole— there was almost no capacity for looking ahead to see what may be upcoming so you could plan in advance. Although it was too late to eliminate the roller on the GT-4 spacecraft, it was deleted on spacecraft 5 through 12.

Young was a bit annoyed by what he called the floating "D" ring. Apparently whoever designed the in-flight storage of the "D" ring handle that, when pulled, would eject both pilots, had not considered storage of the handle once in zero gravity. During the flight the handle kept floating up between his legs and although the safing pins were in the seats, the loose handle was still a safety hazard.

Both crewmen came to the conclusion that some sort of bell, light, horn or all of the above should be installed in the spacecraft to alert them to rapid changes in cabin pressure. Once in the vacuum of space, and after experiencing the primary DC-DC converter failure that sent their cabin oxygen meter to zero, they suddenly realized that if the loss of pressure had been real they would have had no warning.

Grissom set up the "reticule" that would allow future crews to visually sight a nearby docking target and discovered that it really impeded his vision. It also did not seem to line up properly with his line-of-sight. The device was supposed to allow a gloved astronaut to easily fix it in place and have it pre-aligned to sight directly down the nose in the horizontal axis (known as "+Z"). When Gus aligned the nose of the spacecraft with the horizon and had his 8 ball indicating the horizon, he found the pipper on the reticule was "stuck way above the horizon."

Another loser in the eyes of the GT-3 crew was the big plot board that the crew was supposed to carry. It was simply too big and bulky to be useful in the cramped Gemini cockpit. Both astronauts recommended deleting it— NASA never did.

One thing that Young did find to be remarkable was the fact that he did not seem to need his seatbelt and shoulder harnesses. He found that in zero gravity he could simply sit with his head wedged against the hatch and he was held in place by passive friction alone.

Of course picking out these little nits in the spacecraft, its systems, and equipment was what the three orbit GT-3 mission was intended to do.

(MEMORIES OF OTHER FOLKS) Robert; USA: "Way way back in '65, I was able to watch the launch of Gemini 3 and later learn about the 'infamous' Corned Beef Sandwich. That's one of the reasons that John Young is my favorite astronaut. But, I was not able to watch the launch of Gemini 4, or see the pictures from Mariner 9. At that time I was at Fort Dix New Jersey undergoing basic training and for the first four weeks all we were allowed to see was our sergeant's "pretty face." Finally we were given passes to go to the PX (store). When I got there I bought a hot fudge sundae "Yummy" and I bought every magazine that was in the bookrack, back then there were many more magazines then there are today- at least 10. I remember the first picture I saw was that famous photo of Ed White making the first (US) EVA! Beautiful! I also saw for the first time the pictures of Mars taken by Mariner 4. I thought that they were pictures of the Moon— so many craters. Many people forget how important and exciting Gemini was- I'm glad I was able to be there to follow each of the flights- good stuff!"

GEMINI III
AN AUTOMATIC SUPERFINE
RETROFIRE DOWN THE LINE

For the remainder of the mission the crew did platform alignments (resetting the guidance system), suit integrity checks, a high-flow oxygen test and Young did a second demonstration of the urine disposal system that took nearly half of an orbit to complete.

The biggest concern, however was the fact that both of their Flight Director Indicators or "FDIs" were drifting. Young's had tumbled the worst and had been bothering him since two minutes and 51 seconds after liftoff. Grissom's FDI began to act up early in the second orbit and he had to mess with it until the end of the flight. After a while it appeared that when the spacecraft was in orbital mode the FDI ball would begin to drift, but when they were in Blunt End Forward (BEF) or Small End Forward (SEF) the ball no longer showed a drift. That was a good thing, because Gus needed it to help fly the reentry which would be conducted in BEF mode.

Before retro fire the spacecraft's electronic timer triggers a relay and sets in motion the system to arm the separation of the adapter section and retro rockets section of the spacecraft. Located vertically along the right side of the main instrument panel were a series of clear plastic covered buttons; SEP OAMS LINE, SEP ELECTRIC, SEP ADAPT and AUTO RETRO. When the system had been energized those four buttons lit up amber. Normally, the manual called for the commander to push the buttons, but Grissom and Young had decided that Gus would focus on flying MOLLY BROWN and Young would do the switchology. Grissom had already configured the spacecraft into the BEF attitude needed for reentry and he was ready to go. At 04:32:24 into the mission Young called for the switches:

"Separate OAMS?"

"Yeah," Grissom replied, "go ahead."

With that Young uncovered the first button and punched it, then he immediately went down the next two,

"Separate electronics, separate adapter."

There was loud hollow "bang" and the whole spacecraft shook as the SEP ADAPT button turned green. It had activated two redundant linear shaped charges that ran around the attach ring that had been holding the adapter section to the retro section of the spacecraft. As the adapter moved away from the remaining spacecraft a distance of more than one and one half inches it triggered three switches. Those switches in turn illuminated the SEP ADAPT button green and indicated to the crew that the adapter was indeed departing. When asked by the ground if the adapter had separated, Grissom replied that they could feel it go. It was 53 seconds before retro fire.

MOLLY BROWN's onboard timer counted down the seconds until retro fire. Young called for the ARM

AUTO RETRO button and Grissom uncovered the button and punched it simultaneously calling out his action. Now it was all up to MOLLY BROWN.

Below, on the waves of the Pacific, the controller aboard the tracking vessel ROSE KNOT VICTOR counted down the final seconds. The procedure was that at his count of zero the retrorockets would automatically fire. As a backup, Young would punch the MAN RETRO FIRE button one second later just in case the automatic system may fail. There was a maximum 2.0 second delay that could be tolerated in the ignition in order to land anywhere near the splashdown target area.

If the retros did not ignite, Grissom and Young had a back-up to ensure they would return to earth. Back at the 04:21:00 mark in the flight the crew had performed a retrograde burn of 109 seconds duration using the OAMS thrusters. This little burn was reportedly in response to the novel "Marooned" by Martin Caidin. In the original story a Mercury astronaut is stuck in orbit due to a malfunction in his retro package. When some worry warts among the powers that be asked if that could happen on Gemini, NASA's only answer was "Yes." To quench what could easily become a political wild fire, the mission planners came up with an OMS retro fire back-up procedure. This procedure had Grissom firing the OMS thrusters in retrograde at about 04:20:00 into the mission for a 109 second burn. That put MOLLY BROWN into a new orbit whose perigee would intersect on the next revolution with the Earth's atmosphere and cause reentry even if the vehicle's four retro rockets should malfunction. The worry warts were happy and Martin Caidin would have to wait for Apollo before his novel would become a movie.

All Gemini spacecraft each carried a cluster of four solid propellant retro rockets. Each fired for 5.5 seconds in sequence with a slight overlap and each developed 2,500 pounds of thrust.

As the countdown toward retro fire ticked off to the 30-second mark the ARM AUTO RETRO button illuminated and Young punched it, then Grissom reached down next to his left forearm and threw the number "1" RETRO ROCKET SQUIB toggle switch from the "SAFE" to the "ARM" position. He waited a heartbeat to ensure that the retro itself did not fire with that action and then threw the other three squib switches to the "ARM" position. He estimated it took him less than eight seconds to accomplish that task.

"Auto retro armed," Grissom called out, "squibs armed."

CAPCOM counted down to zero.

"Auto retro." Grissom reported calmly as Young waited two seconds and punched the MAN RETRO FIRE button- just in case.

In order, all four retro rockets fired exactly as planned. In fact the fourth retro rocket tried to yaw the spacecraft, which had been programmed into the preflight simulations— Gus was ready for it.

"An automatic superfine retrofire down the line!" Young spouted over the radio.

Jettison of the retro section came within 30 seconds of Retro 4's termination. Again there was the loud "crack" of the LSC as the connection between the two modules was severed with the pushing of the RETRO JETT button. Now, rather than falling around the

Earth in orbit, MOLLY BROWN was falling back to the Earth.

In Mrs. Schroder's second grade class we were heavily involved in afternoon recess as GT-3 reentered.

Since it was well below freezing outside and Nelle Haley Elementary School did not have a gym, we did our recesses inside— at least until the temperature outside got up to freezing.

That day we were probably playing some harmless in-your-seat game such as "Seven Up." That was a game were seven of your classmates are chosen by the teacher to stand in the front of the room and everyone else puts their head down on the desk and closes their eyes while holding up a thumb. Then the seven who were up would each pick some thumb to push down and go back to the front of the class. If you were picked, you got to try and guess who picked you. If you got it right, you took their place at being "up."

I hated that stinking game; I never got picked. To this day I am not sure if it was because the other students knew that I was the most keenly savvy and aware individual in the class and would most certainly know who it was who had picked me, or if it was because I was just a scrawny misfit dork. Thus, while I anguished at my desk wishing for "seven up" to end, the older kids in the school's other classes got to watch the splashdown coverage in Mr. Leady's room.

(MEMORIES OF OTHER FOLKS) Ken; Kansas: I remember Gemini 3 and how our 5th grade teacher dedicated the whole day to us watching the TV coverage and when that wasn't on he got some on the radio. The radio guys sounded like sports announcers for space. When it came time for the splashdown the

TV had it covered, but there was no real, live, pictures. Instead there was just a still picture of an aircraft carrier with some guy talking and trying to tell us what was going on. I remember a lot of confusion about where they thought the Gemini had landed and it took a real long time before we knew that they were okay. When the news came that they had found the Gemini, our teacher put his hands above his head and clapped; we were all very excited.

GEMINI III
LDG ATT

As MOLLY BROWN began to reenter, both crewmen set up their Flight Direction Indicator to give them the critical information needed.

Young's FDI was set to show his rates of attitude deviation, such as pitch and yaw, while Grissom had set his to show cross-range on the vertical needle and down-range on the horizontal needle.

Just like flying a flight director in an aircraft, all Grissom needed to do was to fly toward the needle. The catch was that unlike flying an aircraft, however, Gemini got most of its directional control in reentry by rolling.

The reentry module of Gemini was designed with an off-set center of gravity that could provide some lift during reentry. So, flying it in a "wings level" attitude would actually give the spacecraft a negative lift and allow it to land more to the west, or short of the target point. Flying the spacecraft inverted, however would maximize the lift and cause it to land farther to the east, or beyond the landing point. Thus, Grissom could look at his needles, which were pre-programmed to what was supposed to be his available lift, and if he crossed both needles, he would be directly on target for splashdown— or at least that was the idea.

The problem in all of that planning was that no one knew for certain just how much of that "lift" a reentering Gemini would actually produce. All that there was to tell them what the spacecraft may do was data from Mercury and other assorted reentry shapes as well as wind tunnel data from models. Yet another factor was that solid propellant rocket motors do not shutdown with split second precision; they tend to "tail off." MOLLY BROWN had just fired four large solid rocket motors to break her out of Earth orbit and all four of those motors produced a terminal impulse that could not be accurately pre-calculated. Combined they could have given the spacecraft an unexpected bit of thrust. Both factors came into play as Grissom set up for reentry.

Immediately Gus' FDI showed his down-range needle pegged at the top of the 8-ball and the cross-range needle pegged all the way to the right. He rolled the spacecraft over and held her inverted to gain the maximum lift and as the down-range needle started to come down he rolled out 45 degrees to the right. After about two minutes the down-range needle began to come down and the cross-range needle centered. Then the down-range needle just stopped at about mid-scale, so Grissom rolled back to fully inverted, but the needle bounced a little and found a home right there— mid-scale high. MOLLY BROWN was going to be very short of its target.

Grissom's read of his instruments gave him the conclusion that they were going to splashdown about 50 or 60 miles short of their expected landing point. Indeed, Gemini was producing far less "lift" than the engineers had calculated.

Meanwhile, Young was pretty much ignoring his FDI needles— because there were almost no rates of diversion from the expected course. Anything that did pop up Grissom handled easily and quickly with his hand controller.

So, Young spent some time looking out the window.

He did, however, have an experiment to try. MOLLY BROWN was equipped with a pair of containers in her belly that would spray distilled water into the layer of plasma that surrounded the spacecraft during reentry. Someone had come up with the idea that such an impingement would open a gap in the ionized layer that would then allow radio signals to penetrate the reentry blackout. This experiment required CAPCOM Gordon Cooper to call and count from one to five and five to one continuously during the reentry blackout. Young and Grissom were supposed to listen and see if they could hear him.

It didn't work for voice communications, as the crew heard nothing, but later investigation showed that some C-band and UHF signals coming from the spacecraft were picked up by ground stations.

Unlike Grissom's ballistic reentry on LIBERTY BELL 7, where he was exposed to nearly 12 Gs, MOLLY BROWN only exposed the two astronauts to a maximum of 4 Gs. Yet, since Grissom was holding the maximum lift attitude, they took 4 Gs for the maximum time- all the way down to about 80,000 feet.

Meanwhile, in Mrs. Schroder's class, I had my head down on my desk and my thumb in the air and still— nobody picked me. I began to consider a new mouthwash.

Coming through 50,000 feet MOLLY BROWN

was about to go sub-sonic. The sonic boom would normally be heard aboard the carrier, but this time the prime recovery ship, the USS INTREPID, was probably too far away to hear it.

As the spacecraft plunged through 50,000 feet Grissom hit the DROGUE switch on the center instrument panel near his right knee and deployed the spacecraft's drogue parachute. As soon as the spacecraft stabilized he reached over toward Young's side of the center panel and flipped the two RCS switches, one for ring "A" and one for ring "B" to the center "OFF" position. But as soon as the two rings quit firing the spacecraft began to oscillate under the drogue— so he placed the RCS back on to stabilize the vehicle. As soon as things smoothed out he told Young to switch the RCS back to "OFF." Again the Gemini began to oscillate and Young had to switch the system back on. Before long the barometric switch on the inlet snorkel door activated the system's squib and the door was blown off as planned to let in fresh air. But with the RCS still firing residual fumes from the thrusters began to come into the cabin in small quantities. Both astronauts had their faceplates closed on their helmets and were protected from the toxic exhaust, but the intrusion was not a welcome one.

At 10,600 feet Grissom hit the PARA button next to his right knee on the center panel. That switch released the drogue chute and it in turn pulled out the pilot chute which pulled out the main parachute. In about ten seconds the main chute de-reefed and billowed orange and white in front of the crew. In their heads they counted to ten and then Gus hit the LDG ATT button that was just to the right of the PARA button.

This was switch would drop the spacecraft from a fully nose up attitude to a 45-degree nose up attitude by releasing a second riser. This attitude was considered best for a water impact and was also a derivative of the "Paraglider" or Rogallo Wing flight to landing recovery that had originally been planned for Gemini.

On May 8th, 1961, when the Mercury Mark II program (that later became Gemini) was in its early phases of planning, NASA's Space Task Group (STG) considered that a great deal of time, expense and danger could be avoided if an ocean recovery of the spacecraft and crew could be avoided. Thus they began the study of the Paraglide Landing System. This consisted, originally, of a huge inflatable delta wing contraption that would be stored between the heat shield and the astronaut's compartment. After drogue release the inflatable wing would pop out from behind the heat shield and proceed to deploy by inflating. The Gemini would then, gliding under pilot control, proceed to land on a desert lakebed.

On March 7th, 1962 McDonnell, the prime contractor for the spacecraft submitted their design for deployable landing skids for the Gemini. These three skids, two in the belly and one in the nose, would extend- allowing the spacecraft to land after its paraglide. The main skids and their struts would pop out through doors in the spacecraft's under skin.

Although the paraglide would never fly, those doors were retained on the flight spacecraft. Just over a year after it was proposed, however, the paraglide concept was already falling behind schedule. There were deployment issues with the Rogallo wing and aerodynamic control issues as the system was tested with a ½ scale model of the spacecraft.

By September of 1963 plans were instead being laid for the standard parachute recovery of Gemini. In that plan the two-point sling suspension of the spacecraft was retained from the paraglide concept.

In fact, the Rogallo wing suffered from an aerodynamic flat-spin characteristic that would delay it beyond Project Gemini's ability to make use of it as a recovery system.

Yet, as Gemini was in development testing on the use of the wing an event took place that led to one of the Gemini 3 crew becoming one of the few astronauts to actually fly a Rogallo wing.

That astronaut was Gus Grissom, and he did not have a pleasant flight. The date was October 22nd, 1962 and at Edwards Air Force Base NASA's premier test pilot, Milt Thompson, was conducting a series of test flights on a little aircraft called the "Paresev." This little contraption was designed to research the flight characteristics of flying the Rogallo wing. It had three wheels for landing gear, a small wind screen, a rudder and a control stick— it was also a pain in the butt to fly. Of course Thompson had written procedures that made the darned thing even harder to fly.

The test flights consisted of high-speed tows behind a truck, then low flights up the 20 feet while being towed, then towed flights up to 200 feet and finally flights to high altitudes while being towed by an aircraft. Grissom had done a ground tow on his first day and two days later went for a 20-foot high air tow; the results were not to his liking.

Following Thompson's procedures Gus lifted off, went up to 20 feet, cut loose the tow line and pitched over to get to the flare speed of 55 knots. What he discovered was that unless you pitched over instantly upon releasing the line, you would not have the time to hit 55 knots and flare. Thompson had not told him that, and so he ran out of altitude and as he tried to flare the landing, he hit nose first and snapped the nose wheel off of the Paresev. Thompson got a scolding from an angry and embarrassed astronaut.

As Gemini 3 descended on its parachute and Grissom punched the LDG ATT button both he and Young got another rude surprise. When the risers let go the nose of the spacecraft abruptly dropped! Both astronauts were thrown forward and went face-first into their windows. Grissom's visor hit the reticule knob and cracked while Young's hit something unidentified and was scratched. It was a completely unexpected shock that both astronauts passed on to future crews; watch out for the LDG ATT.

I wonder if Gus sat there happy that he would not have to land MOLLY BROWN by flying it in on that Rogallo wing like he had done on the Paresev.

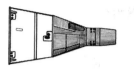

GEMINI III
REGULATION NAVY BATHROBES

Records conflict on just how short of the planned landing point GT-3 splashed down. Some sources say the impact was 60 miles away, some say 50, some say 55 and some say it was 59. To Grissom and Young it made little difference, they had completed a successful mission. Upon splashdown the spacecraft was inverted in the water and Gus saw nothing but blue ocean through his window. For a moment he flashed back to Liberty Bell 7. Then it struck him that they were being dragged along by the parachute. He hit the PARA JETT button that was located next to the LDG ATT button and the aft bridle for the parachute was released. GT-3 was then free and quickly righted herself and bobbed to the surface.

Floating far from the recovery carrier, MOLLY BROWN was sighted by "Big Box 15" a C-54 aircraft that was the military version of the four engine DC-4.

Aboard Big Box 15 were para-rescue swimmers equipped with a flotation collar and life rafts. This was the sort of rescue that these guys lived for. They parachuted down on the landing site and in short order had the flotation collar on the spacecraft. Inside MOLLY BROWN, however, Grissom and Young had already been sweltering in the heat for nearly 30 minutes. They quickly found that their G3C spacesuits were torturous as they allowed so much heat to build up that neither astronaut could stand it.

If they took the suits off all the way it would not allow Grissom or Young to use their head sets and communicate with the rescue aircraft, so both men got their upper bodies out of their suits. Once the rescue swimmers had the flotation collar around MOLLY BROWN, both astronauts got completely out of their suits.

One of their chief recommendations after the flight was to have crews shed their G3C suits as soon as possible after splashdown. No future crews, however, flew in the G3C- all future crews flew in the G4C suit with the exception of the GT-7 crew who flew in the light weight version of the G3C which was dubbed the G5C suit and also Wally Schirra who flew in a G3C on GT-6A.

Fortunately, as Grissom and Young were hoisted from the spacecraft into the helicopter minus their spacesuits, there were two regulation Navy bathrobes waiting for them in the chopper. Thus we have all been spared the sight of seeing them paraded across the carrier deck in their underwear.

Following a phone call from President Johnson and a short flight back to the Cape both Grissom and Young would find themselves cast into the astronaut's version of the pit of human excrement... countless interviews, ticker tape parades and other assorted public appearances. They were indeed national heroes— at least for a few awful months. To them Gemini 3 was just a mission, but to the public of the free world it was an amazing event. It demonstrated that the United States had a new booster, a new spacecraft and had set its sights on the moon.

(MEMORIES OF OTHER FOLKS) George: "Mercury, GEMINI, and Apollo...the three big steps to the Moon. I followed every one of the missions and kept a close eye on the space flight programs with a burning passion. Lots of great stories all along the way!

For example, I was in high school when Gemini 3 was launched. In order to follow the launch of Grissom and Young, I bought a small transistor radio and smuggled it into the classroom. As a result, I heard Jay Barbree (NBC News) reporting the countdown and launch of Gemini 3. For sure, a most exciting and historical day for our country."

GEMINI
A BIT OF CONTEXT

Looking back across NASA's rich history, the modern makers of TV documentaries like to refer to Project Gemini as the over-looked, or over-shadowed, or even the forgotten program. Some like to give the impression that everyone was so looking toward Apollo that no one paid much attention to Gemini.

In fact, nothing could be farther from the truth.

To those of us who were growing up with spaceflight, in the years between Mercury and Apollo, NASA's manned space program simply WAS Gemini. It was the pinnacle of technology; it was THE spacecraft of the United States.

When you thought of the space program, you thought of Gemini. Toy manufacturers loved the Gemini spacecraft because it was shaped a lot like Mercury, so it was easy to take a poorly-molded Mercury and label it as Gemini.

(See Image 5, page 159.) On TV the "I Dream of Jeannie" sitcom, which debuted at 8 p.m. Eastern Time, on Saturday, September 18, 1965, the show's astronauts, Tony Nelson and Roger Healey, flew on Gemini spacecrafts well into the Apollo era. In 1964's movie "Robinson Crusoe on Mars" the escape capsules were Gemini spacecraft with Lunar Module legs attached. If you thought of spacecraft in those days, Gemini was it!

Martin's Titan II booster was the new booster for the Gemini program, and calling it "new" is very much in context. That is due to one commonly over-looked fact about the booster and the Gemini program.

When NASA conceived the prospect of an "advanced Mercury" they were approached by the Martin Company which was then developing the Titan II ICBM weapon system. On May 8th, 1961 Albert Hall of Martin briefed NASA's Associate Administrator Robert Seamans on the possibility of using the Titan II for Gemini, which was then called "Mercury Mark II."

What is important to keep in mind is that at that time no Titan II had yet flown! In fact, the first Titan II would not be test flown until March 1, 1962, some 10 months after that meeting. Additionally, the predecessor of the Titan II, the Titan I was still in developmental testing and would not finish that until January 29, 1962.

On October 27 of 1961 NASA's Space Task Group came up with the plan that would solidify the aspects of Mercury Mark II and it included the Titan II as the booster. That was still more than five months away from that rocket's first test launch. Thus, the entire Gemini program had its foundation based upon a launch vehicle that did not yet exist.

Although those circumstances may sound strange, they were the way that manned spaceflight would progress from that point on. The Apollo lunar program was based upon the Saturn V which was not test flown until November of 1967 and the Space

Shuttle program was started in 1969 and would not see a shuttle flight until 1981.

Additionally, when documentary writers today reflect on Gemini, they sometimes cannot resist the urge to hype-up the drama to events that were dramatic enough on their own. One good example of this is the Discovery Channel's, "When We Left Earth" series. When introducing Gemini and the Titan II booster the narrator says: "…one out of every five Titans fails…" Then they show a series of Titan I missiles exploding.

The tone of the statement has the average viewer thinking that there had only been five Titan II test launches and one failed.

Later the narrator adds: "…Finally NASA launches two (Titan) rockets that don't explode." Thus, implying that every previous rocket had simply exploded.

In fact, the Titan II boosters flew 31 development flights prior to Gemini 1 and exactly one simply blew up on its own. That was the N-7 Titan II and it was the first launch of the vehicle from a Vandenberg silo on February 16, 1963.

Three of the second stage's umbilical lines failed to properly disconnect at liftoff and when they ripped partially lose from the vehicle they damaged the missile's guidance cabling. This left the vehicle in a slow roll. Range safety tried to command destruct the vehicle, but since the umbilicals were still somewhat attached, the missile's onboard logic thought the bird was still on the ground and blocked the signal to the destruct packages. So N-7 just ambled on until it eventually tilted horizontal and the weight of the second stage broke the hull apart. That pre-mature separation triggered the destruct on the first stage but the second stage remained intact and landed in the Pacific.

A second Titan II vehicle, N-20, was in fact destroyed by range safety at 18,000 feet altitude off of Cape Canaveral on May 29, 1963. A fuel leak in one of the engines caused an in-flight fire that damaged the flight control system. The missile slowly began to pitch over and was subsequently destroyed.

In both of these cases any Gemini crew would have been able to eject from the spacecraft as neither rocket just suddenly blew up unexpectedly.

Although there were a total of seven Stage II failures, none were catastrophic and all would have allowed crews to abort safely. Thus, before NASA launched a single Gemini Titan II, the Air Force and the Martin Company had launched 32 Titan IIs, two of which broke up and seven of which simply failed to fully burn their second stages. A total of 23 development flights had been successfully flown.

The actual two birds that "NASA launched," GT-I and GT-II, did not blow up either, but no one had expected them to do so. Through the developmental flight test process the reliability of the Titan II booster had increased to the point where it was dependable and could safely loft astronauts or an atomic warhead. The Gemini astronauts would be flying on as safe and reliable a booster as anyone could want in the mid 1960s. The Titan II was not the implied death-trap that the Discovery Channel would have you believe today.

However, if you look at just the numbers, they were indeed wrong to say that one in every five failed. The average was actually worse- it was one in every 3.5. By the end of the Gemini program, 12 Gemini Titan II vehicles had been launched for NASA; zero had failed.

However, the real hitch in getting Gemini off the ground on the Titan II was more about turf than technology.

The development of the missile began in May of 1960 when the Martin Company signed a contract with the Air Force to create a successor to its current primary ICBM, the Titan I.

Although the Titan I was a far better weapon system than the Atlas, it too had some drawbacks. The biggest of those short-comings was the fact the Titan I used a cryogenic oxidizer that was not storable for long periods. That meant that while the Titan I needed a protracted time period to be prepared for launch, the Titan II was a quick reaction missile. The Titan II would use propellants that could be stored and it would be a key-turn away from launch should the Soviets attack. Thus, it was slated to be the Air Force's "big stick" in the Cold War.

Yet, just a year and a half after signing the contract with the Air Force, Martin was courting NASA as another customer for the Titan II. The Air Force was less than happy with that. In their eyes the Titan II was to be a weapon system and nothing more. Its purpose was to deliver a large nuclear warhead to the Soviet Union, period. But, NASA had some man-rating requirements that would alter the development a bit. What followed was a truly uneasy relationship between the Air Force brass and NASA's upper management.

President Kennedy's challenge to put a man on

the Moon and return him to the Earth before the end of the decade, however, had given NASA a rapidly growing amount of political clout that caused the Air Force brass to grudgingly yield to the space agency's needs.

First on the list of NASA issues that needed to be solved was the tendency for the Titan II to "pogo" or shake vertically like a pogo stick. This was caused by a harmonic vibration between the vehicle and its process of consuming and burning its liquid propellants.

Oddly, the pogo of the Titan II was not a problem for a warhead, but was a huge problem for an onboard crew. Thus, the Martin engineers came up with a "fix" for the pogo issue. It involved the installation of a surge-suppression standpipe in the oxidizer feed line of the first stage as well as increasing the pressurization by 50%.

Those fixes were applied to the flight test launch vehicle N-11, launched on December 6th, 1962, and instead of curing the problem it got worse.

Indeed, the pogo of plus-or-minus 5Gs and were so intense that they caused one of the first stage engines to fail. That, of course, was not going to satisfy NASA.

Next the engineers decided to replace the vehicle's steel oxidizer feed lines with lines made from aluminum. Considering that different materials have different resonant frequencies, they felt that this would do the trick and applied the change to vehicle N-15. When it launched on January 10th, 1963 the pogo was greatly reduced to only plus-or-minus 0.6Gs. Martin engineers, however, also wanted to add mechanical accumulators in the first stage fuel lines to further the pogo suppression. That change would have to be test flown.

At that point Air Force Brigadier General John L McCoy, who was in command of the development of the Titan II had his fill of NASA's man-rating requirements.

He thought the Titan II was good enough as a weapon and he did not want to incur any undue delays by having to wait for Gemini-related systems to be installed.

The General's demand was muted by the fates of flight testing, however, when four out of the next seven Titan II launches suffered failures. He was forced to order a stand-down in testing until everything was fixed.

The two-month stand-down gave engineers the time needed to fix not only the pogo issue but a nagging problem with second stage thrust instability. Vehicle N 25 was launched from the Cape's Complex 15 on November 1st, 1963 and carried the last of the fixes. They worked and the Titan II was ready for Gemini.

Titan Launch Complex 19 had already been planned for space launch use by the Air Force, the last non-spaceflight launch having been a Titan I test launch on January 29, 1962. Thus, it would be the facility that NASA would be allowed to use for Gemini. A lightweight aluminum "whiteroom" was constructed atop the launch vehicle erector and the NASA civilians invaded Complex 19.

GEMINI
THE BARTON FREE-WAY

When you grow up as a space-buff there a lot of modern legends that you repeatedly stumble across. Sometimes they are big and well known and other times they are just tiny and obscure. One that I have always found to be of interest is a little legend that I had often seen mentioned when I read about Project Gemini; it is called the "Barton FREEway."

Air Force Warrant Officer Elmer Barton had been assigned to the Atlas program at the Cape and when Project Gemini moved into Launch Complex 19, Colonel James J. Albert was put in charge of the launch complex. He was wise enough to bring Barton with him.

He would eventually moniker Barton as "The Mayor of Complex 19." Indeed, Barton, who some of the Gemini astronauts describe as reminding them of TV's "Sergeant Bilko" was a person who got things done by hook or by crook. Yet he got things DONE and for the betterment of the program.

Soon Barton became friends with many of the Gemini astronauts. Primary in his circle were Alan Shepard, Pete Conrad, Wally Schirra, John Young, Tom Stafford and especially Gus Grissom.

It was Grissom who brought to Barton's attention

the fact that the crews, who would be suiting up at Complex 16, would have to be transported off of LC-16, onto ICBM Road, down to LC-19, and then across the complex and onto the pad. It was a distance of just a few feet less than one mile. But if there was a single road that ran directly from the LC-16 ready room to the pad at LC-19, it would only be about 1,000 feet.

That short cut would have two great advantages for the crews. First it would mean less time for them to be sealed in their suits before launch and second it would prevent them from having to exit the LC-16 security gate and be exposed to the press. Gus was not at all fond of the press.

He had taken the matter up with NASA but had been turned down due to the cost. Barton told him that the Air Force was also unlikely to take up such a project due to cost, but asked Gus to let him, "Put it on the burner and let it cook for a while."

Of course it did not take long for Barton to figure out a way to scam a road similar to what Grissom and the other astronauts wanted.

As it turned out, an order was issued for crews to "clear the abort area" which just happened to be the land between LC-16 and LC-19. Since Gemini crews would "Pad Abort" by way of the spacecraft's ejection seats and then parachute to safety, they needed a clear area in which to land. Nothing higher than three feet could be left standing and all of the underbrush was to be cleared. After all, it would do no good for an astronaut to eject from a burning Titan II only to end up landing in the bushes and get bit by a rattlesnake. Thus, all of that scrub had to be cleared away. In order to do that clearing trucks and bull dozers and other equipment needed to be brought in and then drive out.

What the mayor of Complex 19 did was to order all of those vehicles to drive the same path in and out while doing the clearing. Before long the crews had blazed a dirt road between the two launch complexes.

But a dirt road soon dries out and all of those vehicles driving upon it can make a lot of dust. Dirt and dust are the sworn enemy of space vehicles. So Barton mentioned to one of the contractors that perhaps they should spray some oil on that road, ya' know; just to keep the dust down. Before long the contractor had a road crew out there asphalting the roadway… just to keep that pesky dust down… and the Barton FREEway was born. Barton took Gus for a drive on the new road and the astronaut was thrilled!

One morning, just before one of the first two Gemini Titan II launches, Barton walked into the LC-19 blockhouse and was met by Alan Shepard,

"You know Chief," Shepard used his Navy vernacular to address the Warrant Officer, "we can't launch today."

"What do you mean we can't launch?" Barton responded, a bit confused.

"No, we can't launch," Shepard went on, "I'm going over to tell Colonel Albert that we can't launch."

"Now come on, Alan," Barton scoffed, sensing one of those famous Shepard gotcha's, "get off my back."

"No," Shepard insisted, "we're in violation of flight rules." Meaning the rules that said nothing taller than three feet was allowed in the abort area.

Shepard then pointed at the blockhouse TV

monitor that was focused on a green road sign stuck on a tall pole. In black letters the sigh read, "Barton FREEway."

Barton was shocked— astronauts Grissom, Schirra and Stafford had gone out and had three signs made and then during the night had planted them along the road! Barton immediately ordered one of his sergeants to go out with some guys and pull the signs down because they actually did violate flight rules. Sergeant Eberle returned a short time later and said that they could not get the signs out of the ground. The astronauts had mounted the steel poles in cement. When Barton dutifully reported the incident to Colonel Albert, his commander told him,

"Just leave them- I think they're marvelous."

At least one of the signs was awarded to Mr. Barton at the end of the Gemini program. He says that he donated it to the Cape Canaveral Air Force Station Space and Missile Museum. It now hangs in the Museum's History Center.

And YES, the Barton FREEway is still there— you can see it on Google Earth.

(See Image 6, page 159.)

(AUTHOR'S NOTE: To moniker Gemini's 1, 2 and 3 NASA officially used the English numerals. Beginning with Gemini IV, however, they changed and officially used Roman numerals— so that is the way it will be depicted in this book. Bet ya' didn't know that, eh?)

GEMINI IV
LEONOV WALKS

At half past four in the morning just five days prior to the launch of Gemini III, I was quietly snoozing in my bed at 3324 Lexington Drive in Saginaw, Michigan as my Dad was shoving railroad cars around the Saginaw yards aboard his C&O switch engine.

It was an overcast, sub freezing night as snow flurries spit horizontally and a blizzard approached from the west. There was no way to look into the sky and see any satellites silently zooming overhead. Yet, up in that black void of space, history was being made as cosmonaut Alexei Leonov opened the hatch on a makeshift airlock and floated freely on an umbilical line outside of the Voskhod II spacecraft. He thus became the first human to walk in space.

Leonov's EVA lasted just over 12 minutes and was the second Soviet spaceflight "shocker" to hit the free world during the gap between Mercury and Gemini, the first being the initial Voskhod mission where three cosmonauts were sent aloft in one spacecraft. The mantra soon began that the Soviets were out-doing the United States, they were out-smarting us, out-engineering us, out-developing and out-tech'ing us. In fact, all they were doing was out-stunting us.

Very little was known in free world about the

Soviet Vostok and Voskhod spacecraft at the time of their flights. It was easy for political and media outlets to make the case that they were far ahead of us; but such was only true until October 27th, 1961 and at that point the Soviets lost their real only lead.

Their advantage had been one of lifting power. Their R-7 booster was the most powerful in the world until that October day when the first Saturn I lifted off from Cape Canaveral's Pad 34. From then on the United States had the most powerful successful boosters.

So far as manned spacecraft, the Soviet program was little more than a series of stunts. The Vostok itself was not an engineered manned spacecraft from the ground up like the American spacecraft. In fact the Vostok was a repurposed Soviet spy satellite called Zenit. Soviet engineers removed the surveillance equipment and installed an ejection seat, some minor instrumentation and a life support system. As with Mercury, the passenger could do little more than ride it up and then down. But unlike Mercury, the Cosmonauts all had to eject and make their final descent under a personal parachute before the capsule landed.

For Voskhod, the engineers stripped the Vostok internally and for the first flight they simply crammed three crewmen inside and shot them into orbit. There was no room for the crew to wear any sort of pressure suits. They also had no form of escape— if something went wrong they would simply die.

Of course the Soviets could claim that they had now sent the first three man crew into space, beating the US planned Apollo spacecraft by three years. The question to be asked, however, would be, "to what end?" Other than propaganda stunting, there was no developmental reason to fly such a mission in such a way. They never did it again, so it served no purpose other than bragging rights.

When it came time for Voskhod II the plan was simple; Leonov was to get out of the capsule and then get back in... period. The launch was again conducted with no escape, but this time the two cosmonauts had on actual spacesuits.

Upon achieving orbit, Leonov did exactly as he was told, he went out, floated a bit and got back in. Days later he and his commander Pavel Belyaev were shown to the world as heroes of the Soviet people.

Many in the west then speculated that the Voskhod was perhaps equal to or more advanced than the Gemini— they were completely wrong. Additionally, no one knew how close the crew of Voskhod II had come to losing their lives. In fact, Alexei Leonov almost became a dead body in orbit.

In order for Leonov to exit the spacecraft, an inflatable airlock had to be attached to the capsule around the hatch. Unlike Gemini, whose hatches, pressurization consumables, and interior were designed from the beginning to depressurize and allow easy egress and ingress, the Voskhod was not.

Some sources say that many of the interior displays and equipment packages were not capable of withstanding vacuum conditions and so the airlock was needed. Others say that the Voskhod did not carry enough consumables to dump the cabin atmosphere and then repressurize it. Still others say that the capsule simply did not have the capability to repressurize. No matter, the airlock was the kluge that solved the problem.

However, once in the vacuum of space, Leonov's spacesuit ballooned to a point where not only were his movements restricted, but he was too large to fit back into the airlock.

"…My feet had pulled away from my boots," he later recalled, "and my fingers away from the gloves attached to my sleeves…"

So deformed and bloated had his suit become that he was simply floating inside of it. That made it nearly impossible for him to get back inside the airlock. Finally he decided to let some of the air pressure out of the suit a little at a time until he could squeeze back into the airlock. Sweating and nearly exhausted he managed to return to the inside of the Voskhod capsule. Apparently, his suit had not been checked in a vacuum chamber, where the ballooning tendency would have been discovered on the ground rather than in orbit.

Once Leonov was safely inside the crew waited until the scheduled time and then they jettisoned the airlock from the side of the capsule.

The problem then became that the explosive bolts that blew away the airlock imposed an acceleration onto the Voskhod, and the vehicle entered into a 17 degree per second roll. The Voskhod was equipped with thrusters that could have stopped the roll, but unlike Gemini which carried a lot of maneuvering system fuel, Voskhod was not provided with enough fuel in its thrusters to use to stop the roll. In fact, they had only enough fuel to do one stabilization burn and that had to be saved for reentry scheduled for 22 hours later.

At length the spacecraft's automatic systems began to crap out and the crew were forced to do a manual reentry. Then the Voskhod's service module did not separate from the reentry module and the whole mess reentered in a violent 10G spin.

Eventually the two came apart and the capsule landed in deep Siberia. They spent two nights in the wilderness before being helicoptered out to civilization. Meanwhile, the Soviet government had lied to their wives and told them that the men were safe and resting in a dacha. The wives were told to write them letters while the Soviet military was still searching for the crew. No other Voskhod ever flew and the next Soviet spacewalk did not take place until January 16th, 1969 when crew members transferred between Soyuz 4 and Soyuz 5.

Thus, what was the point of Leonov's spacewalk? Was it an engineering task to aid in the development of doing work in space? No. Was it a test of equipment and methods for doing rescue in space? No. Was it the first step in a progressing series of EVAs that would further the development of doing EVAs on future Voskhod missions? No. It was done, just like the 3-man Voskhod flight before it, simply to do it. It was primarily a propaganda stunt.

The same can be said for the crew exchange EVA between Soyuz 4 and Soyuz 5. The Apollo 9 mission had been publicly scheduled to do an exchange EVA between the LEM and the CSM. It was no secret and it was not a last minute decision— it was out there in the open that the EVA transfer was in the mission. Then, just two months before Apollo 9, two Soyuz spacecraft, rigged with a crude probe and drogue fixture do the first docking of two manned spacecraft (which was also on the Apollo 9 schedule) and then do an EVA crew transfer. Again; to what end? The answer is simple; upstage the Americans for propaganda purposes. In other words, it was another Soviet stunt.

Leonov's EVA, aside from being an historic first, had one other side effect- it caused NASA to solidify its plans for a free-floating EVA on Gemini IV.

GEMINI IV
PURPOSE AND OBJECTIVE

Long before Leonov's space walk, plans were being made in Gemini for EVA. In fact, the Gemini spacecraft was specifically designed with EVA in mind.

Originally the EVA plan called for the crew of Gemini IV to do the first EVA in the program. A simple stand-up EVA was the intention for GT-4 where the pilot would open the hatch and stand on his seat. As far back as July of 1964, Gemini Deputy Manager Kenneth Kleinknecht stated at the press conference that introduced the crew of Gemini 4 to the public that one of the crew may open the hatch and stick his head out. Oddly, the news media let that remark simply slide past.

The stand-up method was first tried in the altitude chamber with the crew of GT-3. John Young opened the hatch and stood up. All went well except for the fact that the crew had a great deal of difficulty getting the hatch to close and latch. Although the EVA was tested, it was not included in the GT-3 flight plan. Rather it was decided that the stand-up EVA would tentatively be scheduled for GT-4.

When Leonov walked in space a lot of the "powers that be" were appropriately shaken but not stirred by the stunt. Although the task of a free-floating spacewalk similar to Leonov's was still not officially approved, the stand-up EVA was solidified for GT-4. At the time, since no one in the west knew the capabilities or the shortcomings of the Voskhod, there was the real fear that the next Soviet "space first" may be something even more dramatic. Oddly, the truth was that the Soviets were finished for at least the next two years.

Somewhat quietly, the crew of Gemini IV, Command Pilot Jim McDivitt and Pilot Ed White began training for a full free-floating spacewalk.

The new G4C spacesuit which would allow for EVA was already in the final phases of development. Also in development was a cold gas thruster gun that White could use for motion and control when he was outside the spacecraft. Also in the final stages of development was the umbilical that would connect White to the spacecraft.

The crew worked with McDonnell engineers and got their GT-4 spacecraft into the altitude chamber and demonstrated the capability to conduct the EVA. The only hitch was that they too could not get the hatch to lock closed. It was the end of a series of chamber tests that lasted in the neighborhood of a dozen hours. McDivitt and White were tired and did not want to fumble with the locking mechanism, so they simply finished the test in pressurized spacesuits.

Following the end of the test, the crew hit the showers and got something to eat, then McDivitt headed back to the spacecraft where a technician, Carl Stone, was busy working on the locking mechanism. The device had a series of interlocking gears and tumblers that had to meet in just the right way in order for the locking to take place. He watched what the technician did and learned a few lessons that would later save his life.

Yet, the initial purpose and objective of GT-4 was not EVA. Rather the GT-4 mission was to be America's first long endurance mission; it would last four days.

It is important to keep in mind that in 1964 the free world's longest spaceflight had been Gordon Cooper's flight of just 34 hours, 19 minutes and 49 seconds. There were still many in the medical community who thought that two astronauts could not stand the isolation or the effects of weightlessness much longer than Cooper had done.

Some felt that McDivitt and White might actually go insane and perhaps kill one another. Others in the medical community felt that the two might die when the forces of reentry were imposed upon them or perhaps when their spacecraft splashed down their hearts may simply stop due to the impact. So began the medical question of the effect of confinement vs. the effect of weightlessness.

In order to solve that question some of the doctors decided that it would be a good idea to lock the two astronauts inside the simulator for four full days. Of course there was just one small flaw in that hair-brained idea; gravity. The seats in the simulator, like those in the spacecraft were angled 12 degrees off vertical. That meant that when seated upright in 1G the commander was forced to lean to his left and the pilot was forced to lean to his right. The result was that each man was angled over onto one kidney. If the crew were reclined on their backs, the blood in their legs would eventually find difficulty and their legs would "go to sleep" long before four days expired.

It was a problem that, in weightlessness, or when reclined on the launch pad mattered little, but when seated upright in the Earth's gravity, no one could stand it more than about 12 to 14 hours at a time, so four days was insane. Fortunately for the crew of GT-4, the astronaut's doctor, Charles Berry, stepped in and scratched that idea.

Still, Gemini IV's mission would be one of endurance with McDivitt and White venturing into unknown territory of human spaceflight so far as the free world was concerned. Of course Vostok 5 with Soviet Colonel Valeriy Bykovskity spent 4 days, 23 hours and 8 minutes in orbit in June of 1963— but the world outside of the Iron Curtain got exactly zero data from that flight other than the fact that Bykovskity had survived. In fact even the world inside the Soviet Union got little data on what took place during that mission- every fact was smothered in secrecy. Thus, the crew of Gemini IV were not scheduled to beat the record of Vostok 5, they were scheduled to gather medical data- at least in the beginning.

GEMINI IV
ZIP GUN AND STAGE LIGHTS

Following the success of GT-3, NASA management decided to officially establish a point of trivia for future historians. They officially changed the designation of all Gemini flights from a digit, such as "Gemini 4" to a Roman numeral, "Gemini IV." Then they worked at a somewhat harder task of trying to convince one another that a full-floating EVA was what needed to be done on Gemini IV rather than the stand-up EVA.

As the discussions went on through the upper levels of NASA management, the crew of GT-4 trained for both, but put a greater effort into the free-floating version.

Engineers at NASA's Houston Manned Spacecraft Center (MSC) Crew Systems Division had designed and constructed a thing that they called the "hand-held maneuvering unit," also known as the "zip gun." This "T" shaped device had two nozzles on each end of its crossbar and two gas canisters on its vertical beam. The concept being that nitrogen gas would be loaded into the canisters and with the pull of the trigger the astronaut could expel the gas through the nozzles and propel himself in any given direction.

The astronauts spent a good deal of time on a floor that was very similar to a modern air hockey table and using a flat disk they seemed to float as in weightlessness. There was just one hitch- they were operating in one dimension. Thus the gun seemed to work quite well and White gave it rave reviews in his post-flight de-briefing. But, McDivitt later recalled, in the three dimensions of space, "...the gun was practically useless." In order to properly move the astronaut in his desired direction, the gun needed to thrust directly through his exact center of gravity. Quick- where is your exact center of gravity? Don't know? Neither did the astronauts when they left the hatch.

On March 29th, 1965 the top guys at the Crew Systems Division (CSD) briefed the Director of MSC Robert Gilruth on the preparations for the free-floating EVA. They had on hand a mock-up of the chest pack that the astronaut would wear as well as the gun that he would use. Gilruth gave them the go ahead to take the concept to Washington and NASA Headquarters. There they were to give the same briefing to the Associate Administrator of Manned Spaceflight George Mueller. He, however, was not fully onboard with the idea, but advised CSD to keep developing the hardware.

It would not be until May 25th, just two weeks before the launch of GT-4 that the free-floating EVA would be officially approved by NASA Headquarters!

Another last minute addition to the flight plan for GT-4 actually developed from a joke made by Gordon Cooper. While serving as CAPCOM for GT-3, Cooper asked the crew if they wanted to know where their spent second stage may be seen. Gus Grissom replied that they would indeed like to know that information. With that Cooper read off the probable location of the spent stage and then in jest quipped,

"Proceed to see if you can rendezvous."

Of course GT-3 was in no position to attempt any sort of rendezvous, in fact they never even sighted the spent stage. But, listening in the gallery was Robert Gilruth who thought that such a low budget rendezvous may be an excellent thing to try on the next mission. He told George Low about his idea and Low was all for it.

The only snag was that the stage and the spacecraft were not equipped with any sort of rendezvous radar and so any sort of rendezvous would have to be attempted by eyesight alone. To aid in that, the Martin Company attached two lights to the second stage of the GT-4 booster. Additionally, McDivitt and his back-up, Frank Borman, spent a full day in the simulator trying to develop methods for by-the-eye rendezvous. In the end, none of that would help much.

By mid-May the G4C EVA spacesuit that Ed White would wear was finally flight qualified and everything appeared to be in order for the launch of GT-4. The free world's first spacewalk was set to go and the photos and film from that event would forever change the way that people considered the term "astronaut."

GEMINI IV
A TWO BROOM HANDLE LAUNCH

McDivitt and White took their ride up the Barton FREEway from LC-16 and arrived at Pad 19 at 7:07 on the morning of June 3rd, 1965. Every TV and radio network in the United States was covering the event as they stepped from the transfer van and walked up the steel grated access ramp to the elevator. From that moment on, no live picture of the astronauts would be seen and only one "pool camera" was allowed at the pad. That camera, belonging to CBS News transmitted images that were shared by NBC, CBS and the BBC. Now, all of the activity would be funneled through NASA including voice transmissions.

Once sealed inside the spacecraft there was little for the two astronauts to do other than respond to questions from the blockhouse and monitor the active systems. Most of the switchology had already been done by the back-up crew. In their headsets, however, the crew was hearing three different channels, the booster test conductor, mission ops., and CAPCOM. That made for quite a bit of chatter going on and McDivitt would later recommend that the frequencies be cut down to just CAPCOM.

On the other hand their radio suffering was significantly reduced by the fact that in order to hear anything, both astronauts had to turn their volume controls all the way up. Apparently there was an issue with the radios installed in the GT-4 spacecraft. Between themselves the crew wondered if the comm. was that bad 200 feet away, how bad would it be 100 miles into space.

Perhaps the only glitch in the countdown came with the lowering of the erector. At T-35 minutes the blockhouse sent the order to lower the erector, but the big tower just sat there and refused to move. A technician at the pad level found that a relay had refused to close. He fixed the problem by manually closing the circuit and the erector began to lower.

As it moved away from the launch vehicle it caused an unexpected shaking of the vehicle. For a moment McDivitt considered that it just may knock the spacecraft right off of the Titan II and send it crashing to the ground before he ever got the chance to fly it. Suddenly, as the erector got about half way down into its lowered position, it got stuck.

This was nothing new on a Titan pad. The erectors had stuck before, but for assorted reasons. Now, with the whole world watching, what had been an annoyance during top secret development operations, became a major snag in public operations.

When it became clear that the erector would not budge, the fertilizer began to hit the ventilator. A hold in the countdown was called by Robert Adcock, the test conductor, and the guys in the blockhouse who knew the most about the machinery, Frank Carey and Jim Houghton of the Martin Company headed out to the pad.

Meanwhile, in Houston, the ever-impatient flight

director, Chris Kraft began to pester the test conductor about how long it would take to fix the unknown problem. The TC fell back on the standard line that was always used in that sort of situation,

"About 10 minutes flight," Adcock responded.

That 10 minutes soon turned into more than 76 minutes. In his book "Two Into the Blue" Robert Adcock, who was a part of every Gemini launch, provided some insight into the actual cause and solution to the stuck erector. That along with a study of the equipment provided the pieces to the whole puzzle.

Engineers found the problem was in a take-up reel that was supposed to wind up an electrical cable that extended from the erector to the pad. As the erector lowered, the distance between the cable's upper junction and lower junction lessened, but if the cable was not reeled in it could cause some bad stuff to happen to it as well as to the systems that it supported. So, the reel was rigged with a switch that would sense a given amount of slack in the cable. Once that amount was exceeded the switch would trigger and stop the motor that raised and lowered the erector from lowering farther.

As it turned out, there was only a little slack in the cable and the engineers figured that if they could find a way to hold the cable up, the switch would not close. The solution was two broom handles that were cut from their bristles and used by the engineers to hold up the cable. The sluggish up-take reel then did its job and the switch was satisfied- the erector finished lowering. The launch of GT-4 and America's first space walk were again in a "GO" condition- thanks to a pair of engineers with broom handles.

Up in the cockpit McDivitt and White waited and answered what few calls came to them from the blockhouse. When the test of the engine gimbling took place the unexpectedly high level of noise and vibration got the attention of both astronauts.

At the T-30 second mark they felt and heard the huge second stage pre-valves open as the Titan II prepared to come to life.

"Zero! Bolts! Liftoff" came from STONY, astronaut Rusty Schweickart.

Those words plus a gentle push against their backs were the only clues the crew got to allow them to feel that they had left the Earth. McDivitt's event timer had also started as the booster ride became more like a speedy elevator than a jet fighter. Gemini IV was on its way toward space!

In Houston's newly activated Mission Control, Flight Director Chris Kraft watched over his team of controllers intently and one of those controllers, Dr. Charles Berry, watched the heart beats of the crew just as intently.

Unknown to most people in the general public, both Kraft and Berry were also wired up for EKG readings just like the astronauts. It was a part of a bet between the two men in Mission Control to see whose heart rate would be higher during the launch. Berry's heart rate got as high as 90 during the launch, but Kraft's topped out at 135! That was 15 beats per minute higher than Ed White who was aboard the spacecraft atop the Titan II. Just for that, Kraft owed Berry a dinner.

Very little was said by the crew during the ascent.

In fact, CAPCOM Gus Grissom had a great deal of trouble talking to the crew. Through the entire flight there would be a consistent difficulty in the air-ground communications.

Just like the crew of GT-3, the GT-4 astronauts were treated to a high performance ride into orbit. Every system on the Titan II functioned normally.

However just 15 seconds prior to staging the booster began to do a slight pogo. McDivitt felt it, but White did not. Just as the he was considering the pogo, the staging sequence began. Both crewmen found that the staging sequence was far smoother than expected. Ed White later recalled that he did not even hear the second stage's engine start. As the Radio Guidance System (RGS) initiated the vehicle hunted briefly, but hardly changed attitude. Apparently GT-4 was right on target as it accelerated toward orbit. After five minutes and 34 seconds of powered flight the second stage engine cut off and GT-4 was in orbit- thanks to two broom handles, of course.

(MEMORIES OF OTHER FOLKS) Phillip; USA: "The only Gemini memory that I have is because I live on the other side of the Great Atlantic Duck Pond. Thanks to the launch of INTELSAT-1/Early Bird, Gemini 4's launch was the first one to be broadcast "live" to the UK - and presumably anywhere else in Europe that was interested. So BBC-1 broadcast it: it was a day that school finished early for a short holiday so it was the first launch that I saw live on TV."

GEMINI IV
EYEBALL RENDEZVOUS FUN

While the first 30 seconds of orbital flight ticked off, McDivitt unstowed the hand controller. At the half minute mark he called,

"Thrusting. Separate."

With that call White hit the SEP SPCFT switch and fired the linear shaped charges in the launch vehicle mating ring. With a bang that could only be heard inside the spacecraft the adapter was separated from the second stage and the crew began one of the most ill-planned tasks ever performed in NASA's manned space program; the attempt to rendezvous and station keep with the spent second stage of their Titan II booster.

In the world of orbital mechanics normal pilot's flying instincts simply do not work. In fact, that realm of formation flying gets completely turned on its head.

With two vehicles moving on their own individual circular paths at more than 17,000 miles per hour a pilot does not attempt to fly to the other vehicle in order to get closer to it. This is because flying to it would add velocity and cause you to go into a higher orbit and thus allow your target to remain in a lower orbit and actually move away because it has less distance to travel around in its orbit. Thus, thrusting toward the target causes you to go up and away while your target slips farther ahead of you. If you point your nose down and thrust ahead, you still keep falling behind. And that is just one aspect of orbital mechanics that can be misunderstood. If you try and fly around the target you can easily find yourself spending a lot of fuel as your target rapidly drifts away. A neat rendezvous requires extremely precise targeting and very small tiny burns made at precisely the correct points in the orbit. It also requires very complex mathematics, carefully planned far in advance. In the case of Gemini IV none of that was done. McDivitt's orders were simply to "try and station keep with the stage."

At the moment of separation the entire station keeping plan was flushed directly down the toilet by the rules of orbital mechanics. While the spacecraft thrusted forward, the force of the detonating charges that initiated the separation had imparted a retrograde velocity upon the stage that sent it into a slightly lower orbit. Thus immediately the two vehicles were diverging from one another. Then as McDivitt turned the spacecraft around he discovered that not only was the spent stage tumbling, but it was also venting residual fuel. That vent acted like a small thruster and imparted another velocity onto the stage. Thinking that he needed to negate his initial separation thrusting, McDivitt began thrusting toward the stage. Of course since they had no rendezvous radar he had no idea exactly how much thrusting it would take to zero everything out. The hard fact was that from the instant of separation the two vehicles were each in different orbits and there was no way to use just your eyeballs, thrusters and pilot's instincts and get them back together.

Throughout what daylight remained, McDivitt kept pointing the spacecraft at the booster and thrusting and the spent stage kept rapidly moving away. As night quickly fell the crew could see the flashing lights on the booster that had been put there to aid in the rendezvous, but only two lights had been attached and often only one was showing because the stage was rapidly tumbling.

By the next daylight pass the stage was far away and it was clear that they were using a lot of fuel in the effort. McDivitt contacted the ground,

"Guaymas, this is Gemini 4. We're going to have to get a resolution right away whether you want me to really make a major effort to close this last thing or to save fuel," McDivitt asked.

"Gemini 4, Guaymas CAPCOM. Flight advises they like to save the fuel," but then the controller added, "You'll be advised over the Cape."

Once within range of Houston, McDivitt spoke to CAPCOM Gus Grissom,

"I think we ought to knock it off, Gus. It keeps falling, it's probably three or four miles away, and we just can't close on it."

"Right, knock off," Grissom replied, "No more rendezvous with the booster."

Thus the eyeball rendezvous experiment ended in failure after leaving Gemini IV quite a bit short on OAMS fuel. They were just starting the second orbit of a 47 orbit mission and had already spent 50% of their OAMS fuel.

The odd part was that NASA was fully staffed in the orbital mechanics department. In fact one of their third group of astronauts, selected in 1963, Buzz Aldrin had had done his PhD thesis in orbital rendezvous. But because Gilruth's idea to attempt to station keep and perhaps rendezvous with the booster had come so late in the planning for GT-4 that it was essentially an ad-hoc test. No one had the chance to work out the orbital mechanics and no one had given consideration as to what condition that booster stage may be in after separation. Additionally, no one had specified just how long the crew should attempt to station keep with the stage or how much fuel should be used in the process- there were no mission rules for this event. Although McDivitt and Borman had both spent a full day in the simulator trying to find a way to accomplish eyeball rendezvous, they never came near gaining the ability to do it by way of orbital mechanics. So, rather than getting a lesson in how to do station keeping and rendezvous, NASA got a big lesson in how NOT to do those tasks.

(MEMORIES OF OTHER FOLKS) Steve; USA: Space geek since the day I was born (early 1961), I was too young for Gemini III but remember Gemini IV for some reason, and it started my interest in spaceflight. A few years later a Gemini model was one of the first models my dad & I built together and is one of my favorite memories growing up, when Revell rereleased the model (in the 1990s?) I couldn't help buy one and build it again.

Gemini triggered my interest in spaceflight, but it also gives me good memories of times with my late father.

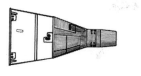

GEMINI IV
EVA AND THE GET BACK IN MYTH

Ed White's EVA was scheduled to begin over the USA toward the end of GT-4's second orbit. There was just one problem- Jim McDivitt had spent the entire first orbit with his eyes glued to that spent booster in a pointless effort to station keep with it. That meant that his full attention was devoted to that task and there was no one to help White don this EVA equipment. In the cramped confines of the spacecraft that was a task that required two people and would take nearly an orbit and a half to complete. So far all McDivitt had been able to do was to unstow the 16mm camera bracket and place the camera on it. The rest of the time he had his eyes out the window and focused on the spent stage and its lights. His major concern was that during the night pass the stage's orbit may bring it back up in close proximity to the spacecraft and cause a collision.

By the time GT-4 came into communications with Carnarvon, Australia, McDivitt informed the ground that they were not going to be ready for the EVA. The flight director agreed. Over Hawaii the official scrub of the EVA for revolution two was made. By the time the spacecraft was over the United States, White was fully suited and basically ready to do the EVA, but McDivitt saw that his pilot was sweating and obviously quite hot from all of the exertion. He was sure that it would do them both good to simply sit still and rest for about 20 minutes.

Before their pass over the USA the crew were alerted by CAPCOM Gus Grissom who told the crew,

"Jim, you're going to be live as you make your pass across the States this time."

That little quip cued the crew that everything they may say on the radio would be going out to the public without a NASA filter.

"Okay," McDivitt replied meekly, "Anything in particular you wanted me to say?"

"Suit yourself." Grissom responded with a note of empathy in his voice.

The fact was that the entire world was expecting the EVA to take place at that time. Of course there would be a spacewalk in another orbit, but no live television of the event, there would only be the voices of the astronauts to keep this exceptionally large audience tuned in. CBS news filled that gap by hiring the famous special-effects artist Peter Foy. He was most well-known for creating the stage flying special effect used in the production of "Peter Pan."

By using piano wire cables that were nearly invisible to the camera in a special harness to be put around the actor, or in this case, a space-suited engineer- Foy was able to make it appear that the person was flying or floating. NBC used a different method for showing their audience what a spacewalking astronaut would be doing. They hired puppeteer Bil Baird and his staff to create three different sizes of astronaut puppets and scale Gemini spacecraft.

Now the TV networks were stuck with their in-studio simulations having actors in spacesuits hanging on wires with a black background and marionettes with their puppet handlers ready to go. Radio stations had cleared the air and had listeners tuned in. So, it was a wise move on the part of NASA to allow live air-to-ground during the pass.

PAO announced it and McDivitt was reasonably chatty during the pass. Describing how the equipment needed for the EVA was now laid out in the cockpit and quipping that they were finally getting the chance to look outside and admire the view. As they passed over Florida and the Cape White told the world that they were looking right down at the launch complexes and could see everything quite clearly. Then both astronauts reported seeing the spent second stage once again, 32 miles ahead of them just as they were going into darkness. Although this seems like very idle chatter to us today, in the summer of 1965 this was huge! The news commentators were universally thrilled at being able to broadcast live astronaut talk as the spacecraft passed over the United States. It very much made up for the delay in the EVA.

Once out of range of Bermuda the crew of GT-4 had rested enough and decided to again go over the full Egress Preparation Checklist from the beginning and recheck every item. Although everything was in place for the EVA, now both astronauts felt sure that they were ready.

When passing over Carnarvon again the crew was given the "GO" for the depressurization of the spacecraft. They lowered the cabin pressure without any trouble.

However, their first indication of a problem with the hatch came when White attempted to engage the opening handle. The unlocking mechanism refused to catch in the gearing- the handle just rotated freely. Having seen this in the altitude chamber, White took his finger and inserted it into the mechanism and moved the gain pawl up until it began to work, from there on he released the hatch.

The timeline for White's EVA called for him to first stick his head out, then stand up and when ready, float out into space. But the depressurization "GO" was actually to be given over Hawaii rather than Australia. McDivitt was wondering why mission control had given the word so early. The answer was simple- Flight Director Kraft had reasoned that the extra orbit had allowed the Earth to rotate a full 90 minutes under the crew and thus the night terminator over the Atlantic was that much closer to Gemini IV. By getting White up and out over Australia it helped to make up for that distance and thus gave him the maximum time outside while allowing him to get back in during daylight.

As GT-4 passed into range of Hawaii, White was standing in his seat and preparing to float free. On his zip gun he had a Zeiss Contarex 35mm camera. Hawaii CAPCOM passed along Houston's "GO" to float free but White was busy with his thermal gloves. They were difficult to get on completely over his suit gloves and before he was able to communicate clearly that he was free of the spacecraft, Hawaii lost their signal. But in McDivitt's final words prior to LOS he seemed to say "He's egressin' right now" But that was not the way that Houston heard it. Now Houston and the entire free world would have to wait for California to acquire the spacecraft- It would be a long three and one half minutes.

Finally CAPCOM Gus Grissom was able to contact GT-4,

"...Houston CAPCOM, Has he egressed?"

"He's out Gus and it's really nifty!" McDivitt gleefully responded.

Knowing that the world would be listening, Ed White said that he thought about what he should say during his EVA. He decided that he would simply perform as a test pilot working with another test pilot.

(See Image 7, page 160.) Frankly, the communications were a bit hard to hear and in all of my years growing up with spaceflight I had always wondered what all was said. Aside from a few NASA voice clips replayed selectively during documentaries, it takes some real digging to get the exact conversation. So, for this book, I have transcribed the entire EVA- because if it was fun and fascinating for me to read, it'll be fun and fascinating for you to read.

What follows is taken directly from the NASA documentation, however, CAPCOM's 32 calls to the spacecraft have been removed. Additionally, upon careful listening to the actual air-to-ground voice recordings, I discovered that there are some omissions in the transcripts, and so I have filled in those areas with what is on the air-to-ground recordings.

McDivitt: Let me try VOX again. Say something.
White: Hello, Gus. How do you read, CAPCOM?
McDivitt: Hey, Gus, do you read?
McDivitt: I read you now, Ed.
McDivitt: It's not really working, Ed.
White: All right.

McDivitt: Okay. Listen, I still can't see you out there.

CAPCOM: It's working. Keep talking.

White: All right. I've drifted off to the side. I think we're going to have ... I'm taking pictures now, Jim.

McDivitt: Okay.

McDivitt: Make that flag look pretty.

White: Yes.

White: Okay. I'm right by the stub antenna now.

McDivitt: Okay. Let me get some ... to you, Ed. I ought to be getting some tremendous pictures of you. Let me try again with the Hasselblad.

White: Okay. I think I've exhausted my (zipgun) air now.

McDivitt: Okay.

McDivitt: Stay right there.

White: I had very good control with it. I just needed more air.

McDivitt: Okay. Stand by. Let me take a couple of pictures, ole buddy.

White: All right.

CAPCOM: Gemini 4, Houston CAPCOM.

McDivitt: Go ahead, Gus.

CAPCOM: Tell him to keep talking. We're receiving Ed through your VOX.

White: Are you reading me, CAPCOM?

White: ... Gus?

McDivitt: Okay. We're going to VOX now.

McDivitt: Okay, Ed. They're receiving you.

White: Okay.

McDivitt: Tell them what you think.

White: That's right, CAPCOM, it's very easy to maneuver with the gun. The only problem I have is that I haven't got enough fuel. I've exhausted the fuel now and I was able to maneuver myself around the front of the spacecraft, back, and maneuver right up to the top of the adapter. Just about … came back into Jim's view. The only thing I wish is that I had more. This is the greatest experience I've … it's just tremendous!

Right now I'm standing on my head, and I'm looking right down, and it looks like we're coming up on the coast of California. I'm going into a slow rotation to the right. There is absolutely no disorientation associated with it.

McDivitt: One thing about it, when Ed gets out there and starts wiggling around, it sure makes the spacecraft tough to control.

White: I feel just about like a … commercial.

CAPCOM: Is he taking pictures? Take some pictures.

White: Okay. I'm going to work on getting some pictures, Jim.

McDivitt: Okay. Get out in front where I can see you again.

White: Yes.

McDivitt: I've only got about three on the Hasselblad.

White: Okay.

McDivitt: Where are you?

White: Right out in front now. I don't have the control I had any more without that gun.

McDivitt: Yes, I noticed that.

White: On the tether.

CAPCOM: You've got about 5 minutes.

White: But I want to get out and shoot some good pictures. I'm not satisfied with that. Okay, I'm drifting down underneath the spacecraft.

McDivitt: Okay. I'm going to start firing the thrusters now.

White: All right. There's no difficulty in recontacting the spacecraft. It's all very soft, particularly as long as you move nice and slow. I'm very thankful to have the experience. It's great, Gus. Right now I'm right on top of the spacecraft — just above Jim's window. I'll bring myself in and put myself out into your view, Jim.

McDivitt: Okay. Move slowly, I'll take your picture.

White: Right now I could maneuver much better if I didn't have the gun with the camera on it, because I have to tie one hand up with it.

McDivitt: Okay. Stay right there if you can. Okay, do you want me to maneuver for you now, Ed?

White: No, I think you're doing fine. What I'd like to do is get all the way out, Jim, and get a picture of the whole spacecraft. I don't seem to be doing that.

McDivitt: Yes, I noticed that. You can't seem to get far enough away.

White: No.

McDivitt: Where are you now? Am I clear to thrust a little bit?

White: No, don't thrust now.

McDivitt: Okay. Ed, just free-float around. Right now we're pointing just about straight down at the ground.

White: Okay, now I'm taking a look back at the adapter and equipment back there. I can see the separation plane; it's quite clean. The thrusters are clean. The thermal paint, the thermal stripping looks quite good. Also, the velcro that we put on seems to be in good shape right by the camera. I'm coming back down on the spacecraft. I can sit up here and see the whole California coast… Okay. Now I'm going to go out and see how much … if I've got enough harness. The sky sure is black.

McDivitt: Hey, Ed, can you see into the right-hand hatch, right now?

White: No.

McDivitt: Aw, shoot!

White: Let me work on back. Listen, it's all the difference in the world with this gun. When that gun was working, I was maneuvering all around.

White: Yes. Okay. … contacted.

McDivitt: Yes.

White: How are you doing old buddy?

McDivitt: Pretty good. How about you?

White: Good. Looking right in your window.

McDivitt: Where? You're not even there. Are you there, Ed?

White: No, I'm moving out now.

McDivitt: Let me take a picture.

White: Okay, there I go. No. Boy, I sure could use that gun.

McDivitt: A pretty neat little contraption, huh?

White: Yes. All the strings that are attached to me tend to dampen out my travel on the lanyard.

(At this point, CAPCOM begins a series of 32 calls up to the spacecraft that the crew cannot hear while they are on VOX- those calls are removed from the text… because they're somewhat annoying)

CAPCOM: Gemini 4, Houston CAPCOM.

McDivitt: Yes, I noticed that.

White: Okay, I'm going to free drift a little bit, and see if I can drift into some good picture-taking position.

McDivitt: Okay. Here, let me control the spacecraft …Just for your information Ed, we're only down to 48 percent on our O2.

White: Okay.

McDivitt: ECS O2 pressure is about 830, so it's staying right up there.

White: You know one thing about this. Where we have this tether attached, I can't get a good push off.

McDivitt: Oh yes, is that right? Where it's attached to you or where it's attached to the spacecraft?

White: Where it's attached to the spacecraft. There's nothing particularly to get a push off on. Now, let's see if I can get a decent push off. If I didn't have the gun I could do a little better. There just isn't anything to push on.

McDivitt: Ed, smile.

White: I'm looking right down your gun barrel. All right.

McDivitt: Let me take a close up picture there.

White: Okay. Just a minute.

McDivitt: You smeared up my windshield, you dirty dog! (laughing)

White: Did I really?

McDivitt: Yes.

White: Well, hand me out a Kleenex and I'll clean it.

McDivitt: Ha! See how it's all smeared up there?

White: Yes.

McDivitt: It looks like there is a coating on the outside and you've rubbed it off.

White: Yes.

McDivitt: That's exactly what you've done.

White: Okay.

McDivitt: Okay. Right now, the spacecraft rates are

up about 1 deg/sec in pitch and yaw, and about ½ deg/sec in roll.

White: I'm not satisfied with the pictures that I've gotten yet.

McDivitt: Okay… We've been tumbling around. I (don't) even know exactly where we are, but it looks like we're about over Texas again… As a matter of fact, you know that looks like Houston down below us.

White: I believe it is, Jim.

McDivitt: Sure is, I — hey, Gus, I don't know if you can read, but we're right over Houston.

White: That's right. I'm looking right down on Houston.

McDivitt: Run out and look!

White: My golly! I'm looking right in the bay there.

McDivitt: Yeah, that's Galveston Bay right there.

White: Yes.

McDivitt: Ed, can you see in your side of the spacecraft?

White: Let me get a picture of it.

McDivitt: Can you see the camera here?

White: Yep.

McDivitt: Is it pointing at you?

White: No, not now. I'm out of it.

McDivitt: Which way?

White: I'm behind the adapter. I'm trying to get some pictures. I'm right above my hatch now.

McDivitt: Okay.

White: Yes, sir. If I had about another couple of bottles, Jim…

McDivitt: That would really be it, huh?

White: Yes. I'm still not satisfied with the pictures I'm getting. I've only gotten about three or four.

McDivitt: Is that right? I've taken a lot, but they're not very good. You're in too close for most of them.

White: Okay.

McDivitt: I finally put the focus down to about 8 feet or so. Oh yeah, that's great, the clouds on water behind you.

White: Okay. I'm going to let myself go out now.

McDivitt: You know, Ed, this thing about the reference we were talking about looks like it's sure right.

White: Out here you don't even need one.

McDivitt: Yes.

McDivitt: I'm going out to PUSH-TO-TALK and see what the Flight Director has got to say.

(With the night terminator rapidly approaching Flight Director Chris Kraft was highly concerned that the crew had lost track of time and Ed White may find himself floating around in absolute darkness. In fact Kraft was about ready to give birth to a bouncing baby hand grenade. He, for the first and only time broke his own rule and spoke directly to the spacecraft.)

Kraft (direct to the spacecraft): The flight director says get back in!

McDivitt: Gus, this is Jim. Got any message for us?

CAPCOM: Gemini 4, get back in!

McDivitt: Okay.

White: One, what are we over now, Jim?

McDivitt: I don't know. We're coming over the West area, and they want you to come back in now.

White: Back in?

McDivitt: Back in.

CAPCOM: Roger, we've been trying to talk to you for awhile here.

White: Aw, Cape, let me just find a few pictures.

McDivitt: No, back in. Come on.

White: Coming in. Listen, you could almost not drag me in, but I'm coming.

McDivitt: You still have three and a half more days to go, buddy.

White: I know it. I'm coming.

McDivitt: Okay.

CAPCOM: You have about 4 minutes till Bermuda LOS, Jim.

CAPCOM: Gemini 4, Houston CAPCOM.

White: I'm kind of maneuvering (garbled)

McDivitt: Okay, okay. Don't wear yourself out now. Just come on in... (19 second pause) How are you doing there?

White: I'm doing great.

McDivitt: Good. ECS O2 ... is back in auto.

White: All right.

McDivitt: Lots of pressure ... 47 pounds ...

White: Roger.

McDivitt: The spacecraft really looks like it's out-gassing because all the — whenever a little piece of dirt or something goes by it always heads for the door and goes on out... Okay, let's — oops, take it easy now. Take it easy.

White: ... I'm standing on top of it, right now.

McDivitt: Okay, ready on top. Come on in.... Boy, this load on our spacecraft is fantastic. You can put in a 2 deg/sec attitude change in just nothing flat!... Can't you get a hold there, Ed?

White: Yeah, go ahead that's my (garbled) gun now.

McDivitt: Okay.

White: It's no sweat. Actually, I'm trying to get a

better picture.

McDivitt: No, come on in.

White: I'm trying to get a picture of the spacecraft now.

McDivitt: Ed, come on in here!

White: All right. Let me fold the camera and put the gun up.

CAPCOM Gemin 4, the Cape, err, you read uhh, Houston now?

White: Ask them about No. 8.

McDivitt: What?

White: Ask them about No. 8. No. 8? Okay, how's no. 8?

McDivitt: Okay. Let's not lose that camera now. I don't quite have it. A little bit more, but easy, easy, easy.

White: Got it?

McDivitt: Okay, I've got lt.

White: Okay, now I can enter … It's just like we said …

McDivitt: Yes, we sort of talked about that. But we sort of talked about that but they didn't have any place for the camera. Come on, let's get back here before it gets dark.

White: Okay. This is the saddest moment of my life.

McDivitt: (Laughing) Well, you're going to find a sadder one when we have to come down from this whole thing.

White: I'm coming.

McDivitt: Okay.

White: It's been an (garbled) just great.

McDivitt: Come on now… Okay.

CAPCOM Gemini 4, Houston CAPCOM.

McDivitt: Be sure you've got those hatch dogs fixed

now.

White: Yes …

CAPCOM: Gemini 4, Houston CAPCOM. Gemini 4, Houston CAPCOM. Gemini 4, Houston CAPCOM.

McDivitt: I just took all that stuff down there and … Okay, I'm going to PUSH-TO-TALK now. I couldn't …interphone. Let's see what the flight director has to say. Okay, have any messages for us, Houston?

Kraft (direct to spacecraft): Yeah! Get back in!

CAPCOM: Roger yer' on your way to get him… Yer' getting him back in?

McDivitt: Yeah, he's standing in the seat now. His legs are down below the instrument panel.

CAPCOM: Okay. Get him back in. You are going to have Bermuda LOS in about 20 seconds.

McDivitt: Yes, he's coming in. He's having some trouble getting back into the spacecraft it looks like.

CAPCOM: You got your cabin lights up bright in case you hit darkness?

McDivitt: I can't read you, say again.

CAPCOM: Gemini 4, Houston CAPCOM.

CAPCOM: Gemini 4, Houston CAPCOM

McDivitt: Go ahead, this is Gemini 4.

CAPCOM: Roger. Is he getting back in, Jim?

McDivitt: Listen, we are kind of busy. If you don't really have something for us, wait a couple of seconds.

CAPCOM: Roger.

With that GT-4 went out of range for the a bit more than the next 23 minutes; the EVA was over.

From reading this transcription it becomes quite clear that as soon as McDivitt told White to come back into the spacecraft, the space walker began to do so. He did not disobey the order. Additionally, the crew was not ignoring Mission Control, a fault in their VOX system simply cut off the transmissions from Houston. PAO, in their post-EVA announcement pointed out this fault and in the press conference following the EVA, Kraft fully explained the fault to the news media. But the media would not take the answer. One reporter forcefully asked Kraft,

"Is it likely that these men will receive an official reprimand for their enthusiasm that they displayed in the extra vehicular moment there? Do y'all plan to sort of rap their knuckles for this?"

"Absolutely not!" Kraft shot back without a heartbeat of hesitation.

A second reporter asked if White may have not wanted to come back in because he was suffering from a narcosis similar to what deep divers sometimes experience. Again the answer was "No" and Kraft went on to again explain the VOX issue. But as time went on the truthful explanations were nearly obliterated by the fiction of a crew gone wild. One modern documentary went as far as to find the words "no" and "ont-uh" spoken by McDivitt sometime during the flight and then dubbed it into the voice track so it appears as if immediately after Kraft calls out his "... get back in!" order, it sounds like White refuses. That concocted refusal is not in the transcripts nor is it on the tapes.

What is easily overlooked is that White needed to get back into the open hatch without any hand-holds other than the spacecraft itself and his tether. That activity alone took a good deal of time. Granted he wanted to take a few more photos, but he was indeed on his way in. McDivitt was coaching him to get back in because White was not in a position where his commander could see him. It is important to keep in mind that McDivitt's field of view was highly limited and so the best thing that he could do was coach White over the intercom until he saw the space walker's feet appear in the open hatch. Finally, it is most important to keep in mind that Ed White was a West Point man- he would NEVER disobey an order. Thus, when you see a documentary, or read a book that implies that the crew of Gemini IV was somehow rebelling, ignoring or disobeying the orders from the ground to "get back in" it is a myth and worse yet- it is poor history.

Getting White all the way into his seat along with all of his EVA equipment was a true challenge. Oddly, prior to the EVA I watched ABC News' Jules Bergman explain with a Gemini model and it's tiny astronaut how McDivitt could rescue White if he somehow broke loose from the spacecraft. He showed how the spacecraft could swoop under White and then just scoop him back into his seat. Of course that was totally impossible, but to a second grader it was pretty cool.

In reality the problem came up that White was fairly tall for an astronaut- his helmet actually contacted the hatch in 1G when he was tightly strapped down. Now in zero G he did not have gravity to help squish him down and with his suit inflated he was even taller. Thus, when he went to sit back down, he actually could not sit! McDivitt did he best to pull White down into his seat, but what little leverage he could apply was simply not enough.

Finally White got his right hand under the circuit breaker panel and pulled for all he was worth and pulling with his legs as well. Then, just as he started to get down he got a cramp in the back of each thigh and had to stand up again and straighten out. At that point McDivitt suggested that he should dispose of his gold visor and White tossed it overboard. On the second try he again struggled and with McDivitt's help finally got into his seat.

The biggest challenge the GT-4 crew now had was one that was literally a matter of life and death. Once White, the umbilical, the zip gun and all of the camera equipment were back inside the tiny cockpit, the crew discovered that the hatch refused to close and latch. Without it closed and locked they could not reenter the atmosphere. They were in a 89.2 by 155 mile orbit with a lifetime of 4.8 days- so they had that long to close and lock that hatch door. Of course it did not take them that long, but it sure felt like it.

Fortunately, GT-4 was out of radio contact the entire time that the crew struggled with the hatch. The problem was the same as it had been when trying to open the hatch. McDivitt, White and technician Carl Stone had disassembled the mechanism at the McDonnell plant and they found that the dry lube that was used in the spring tended to coagulate and cause the spring to stick. They found that McDivitt could use his finger in place of the spring and make the latch work. In space, however, it took the fingers of the gloved hands of both crewmen to finally get the mechanism to finally catch and lock.

Both astronauts were exhausted and dripping with sweat. The effort to get Ed White back into his seat and then close the hatch had been monumental.

White's visor was fogged over as the exertion had overwhelmed his suit's ability to handle the extra heating from his body. Oddly, this high level of exertion from doing relatively simple tasks in zero gravity was a lesson that NASA did not learn and because of that future Gemini space walkers would pay the price.

It would take the rest of the workday to get the leftover equipment half way stowed. Originally, the crew was supposed to close the hatch, put White back on the spacecraft suit system and then re-open the hatch and toss the umbilical, the gun and assorted other stuff overboard. But considering the trouble that they had with the hatch, McDivitt made the command decision to not do the equipment jettison. They would just have to live with the junk for the next four days.

(MEMORIES OF OTHER FOLKS) Doug; Minneapolis: "When NBC News got hold of the 16mm film of Ed White's spacewalk, they rushed it ("still wet from the developers") onto the air. For the first and only time in my memory, they prefaced the EVA film footage with a four-panel version of the NBC color peacock. It was new, it was unexpected, and I still recall it to this day.

The EVA footage, of course, blew me away. I was nine years old."

GEMINI IV
FOUR DAYS AS LAB RATS

Almost completely overshadowed by Ed White's spacewalk was the fact the remainder of the GT-4 mission was actually supposed to be a medical flight and a spaceflight endurance test. In fact, it was the first of three scheduled long-duration Gemini spaceflights. The long-duration flight of GT-4, which lasted four days, was supposed to be followed by an eight day mission and a 14 day mission. Approaching long-duration spaceflight in steps was the objective. Considering that no human being had been in space for more than 4.9 days and no American had been in space for longer than 1.4 days and that no one in NASA could trust Soviet spaceflight claims, the question remained as to whether or not a crew could live in the space environment for any given length of time.

With every bodily function— right down to exactly how many swallows of water each crewman ingested— being closely monitored by a small army of medical doctors, McDivitt and White circled the earth. Meanwhile, on the ground millions of people also kept tabs on the mission, readily soaking up every tiny bit of information the space agency may toss into the public shark tank. Gemini IV was captivating.

The length of the mission was actually determined by two factors, first- the development of the fuel cells that would allow future flights to extend beyond the life of the Gemini spacecraft's batteries and second— the fact that the astronaut's physician Dr. Berry insisted that seven days was simply too long. He felt that some unknown medical issue may suddenly develop and cause a scrub, thus a four-day mission was a sensible step prior to committing astronauts to an extended flight.

Factors such as bone demineralization were thought to be a real threat to astronauts in prolonged weightlessness and in the early 1960s, "prolonged" meant more than a single day. Although it seems silly more than a half century later, the doctors actually felt that human bones might rapidly decay when out of the Earth's full gravity field.

So, one of the medical experiments involved having the pinky finger on the right hand of each of the GT-4 astronauts X-rayed prior to the mission and then having them X-rayed again immediately after returning to the carrier upon recovery in order to see if there was any bone demineralization.

Additionally, the two astronaut lab rats each wore a microphone that was attached to their bio-medical harness under their spacesuit. Although they were already wired for electrocardiogram capability, the microphone's phonocardiogram allowed the doctors on the ground to simultaneously and remotely place a sort of stethoscope onto each crewmember and better determine the interval between heart contractions.

Exercise in the cramped confines of the Gemini spacecraft would seem to be nearly impossible and both McDivitt and White were very much into physical fitness, so finding a way to exercise was a problem. It was supposed to be solved, in a small way, by the M-3 exerciser. This was little more than an elastic bungee with handles on each end. It took a 60 pound force to fully extend it and the flight plan called for the crew to make use of it on eight different occasions during the mission.

White, who was always thought as being the most athletic of all the astronauts later stated that once in the weightless environment he oddly really did not have the desire to do strenuous physical exercise. Both crewmen, however, did use the M-3 as scheduled.

One significant medical item that was gained from the flight, other than the fact the crew's bones did not dissolve and both men did not actually die from their exposure to weightlessness, was the fact they really did not sleep nearly as much as required by the flight plan. While one crewman was scheduled to sleep, the other was supposed to remain awake and working. The problem was that the activity of taking pictures of the Earth and doing star sightings made so much racket in the confines of the cockpit that it kept you awake. It was a simple lesson that would have to be re-learned on Apollos 7 and 8 before the crews were finally allowed to all sleep at the same time.

Perhaps the greatest problem that the GT-4 crew had to deal with was their flight plan. While most of us on the outside of the program picture astronauts as preparing for the smallest detail in their mission for months and months prior to the mission, the crew of Gemini IV had the core of their mission's flight plan handed to them as a final product the night before the launch! McDivitt stated in his post flight debriefing,

"We didn't receive the (final) checklist and the procedures and data books to look at until 8:00 the night before the flight."

He also stated in that same debriefing that when it came to the core of the flight plan NASA,

"...kept changing it around, and changing it around and changing it around."

He explained,

"I think the situation that we finally ended up in was that we tried to have a very good launch in a couple of orbits and a good last orbit or two. All the stuff in the middle, we knew we were going to have to handle in real-time. I finally found myself forced into the situation where I couldn't even worry about the middle flight plan. We had to lay it out in general and expected great number of changes. This is exactly what we did. We handled it in real time."

GT-4 was also the first use of NASA's new Manned Spaceflight Center (MSC) in Houston, Texas and the duration of the mission was such that ground controllers now had to work in shifts and thus the control personnel were broken up into teams. Now all of the data from the mission was funneled down into the soon-to-be-famous Mission Operations Control Room, or "MOCR" which was pronounced "Moe-curr" and known to us who were growing up with spaceflight simply as Mission Control.

So it was that the lab rats in Gemini IV went for 62 orbits handling most of it in real time.

GEMINI IV
ROLLING REENTRY

For most of the mission Gemini IV had been in a conservation mode— they were trying to conserve OAMS fuel and battery power. Thus when a computer update was needed the crew would have to turn on the computer, ground would transmit the update and then the crew would turn the computer back off again. The system worked quite well until 75:42:47 mission elapsed time, and then the fun began.

McDivitt was awake and White was in his scheduled rest period and lightly sleeping when CAPCOM called up and asked for the computer to be turned on so that an update could be received. McDivitt dutifully turned on the computer, ground transmitted up the information and said that their instrumentation showed that the update had been received.

When McDivitt turned the computer switch back to the off position, however, the computer light stayed on and the "Malfunction" light illuminated. He informed the ground of the issue, but their mirror computer showed the computer switch "On" and no "Malfunction" light.

McDivitt tried to troubleshoot by recycling the switch, but got the same results. Then he "recycled it a bunch of times" as he reported to Houston and still got the same results. From there on McDivitt and Mission Control went through assorted switchology and circuit breaker recycling in an effort to fix what was clearly a dead computer. With reentry in the near future, the whole situation did not look friendly.

Normally Gemini astronauts practiced what they called a 90-90 reentry for computer failure situations. That would be that as the spacecraft began to encounter the atmosphere at 400,000 feet the command pilot would impose a 90 degree roll beginning heads down and hold it for a set time, then roll another 90 degrees in the same direction and hold it again. This procedure would be continued through the entire reentry until the spacecraft reached drogue chute deployment altitude. The method was designed to null out any flight path dispersions and was a technique that the crews practiced in training repeatedly. Now, when the actual situation appeared, Mission Control stepped in and made another of those last minute ad hoc changes that the crew had to tolerate.

At 85:35:36 mission elapsed time over the tracking ship ROSE KNOT VICTOR, CAPCOM Jim Lovell radioed up that instead of the 90-90 reentry the crew had trained for, they were now going to have to do a rolling reentry like the Gemini 2 unmanned spacecraft. Mission Control felt that this would cause far fewer dispersions in the flight path.

Four and one half hours later, while McDivitt and White were still probably shaking their heads over that one, Mission Control added some additional ad hoc changes. Now rather than beginning the roll at 400,000 feet, McDivitt would roll the spacecraft inverted at retrofire and then at illumination of the

.05G light he would begin the roll and continue rolling at 15 degrees per second until an altitude of 100,000 feet was reached, then he would roll inverted again in order to obtain maximum lift.

"Shades of Mercury Jim," Lovell quipped sarcastically perhaps referring to the ad hoc decision made by NASA management to leave the retro pack attached to John Glenn's FRIENDSHIP 7 capsule in answer to a false indication of his landing bag being deployed.

"I was going to say I'll bet you Flight Directors were thinking of this one," McDivitt replied, "that's why I say not very imaginative, but I guess he gets the job done."

"Roger," Lovell responded, "I have a bet with your wife that you will get within 5 miles. Now you got to hold up my end here."

"Okay," McDivitt went on, "I'm going to try to hold up my end too. Remember when I said we'd never make a closed looper?"

"Doesn't look like will make one this time." Lovell answered with resignation.

Preparing for reentry, the crew was requested to radio down the types and locations of every scrap of material that they had in the cockpit so that Mission Control could calculate a good center of gravity for the spacecraft. Along the way they were told that the Flight Director wanted to know what the total weight of their "refuse" was?

"About 3,000 pounds!" McDivitt shot back in reply to the absurdity of the request.

Reentry was uneventful considering it was done using an unpracticed procedure. GT-4 splashed down about 50 miles short of its target and Jim Lovell lost his bet with Patricia McDivitt. Likewise the worried doctors also lost their bets when both McDivitt and White failed to die from their exposure to the space environment and to make matters worse, their bones did not dissolve. In fact their pinky finger post flight X-rays showed no difference at all after the four day flight— no matter how hard you studied them.

At best, Gemini IV provided the EVA images that showed the world the true glory and adventure of spaceflight. The photos and films of Ed White floating with the blue earth so far below inspired the free world.

A few years later my folks bought my baby brother a metal globe that had a gray plastic Ed White astronaut on a wire high above the rotating planet. Not knowing the difference between a globe and a ball the toddler soon bashed the globe on the floor and made it into junk. Before my Mom could toss it into the trash, however, I nabbed the astronaut figure and added it to my "stuff." In later years I painted him white with a gold visor and he took many spacewalks from my Gemini spacecraft model.

(MEMORIES OF OTHER FOLKS) Gary; Minnesota: "I met the Gemini 4 crew when they visited my home town, St. Paul, Minnesota, after their flight. I ambushed them in their hotel elevator.

Got Ed White's autograph and talked to him about his hand-held cold gas thruster, which he told me had quarter-pound of thrust.

I was 15."

GEMINI V
EIGHT DAYS IN A GARBAGE CAN

Passing over the Carnavon, Australia tracking station Gemini V established communications. Astronauts Gordon Cooper and Pete Conrad were just an orbit and a half from reentry and finishing a grueling eight-day mission. The ground was concerned because the propellants in their OAMS system were reading zero— in fact, some of the thrusters had just been spitting out liquid for quite a while anyhow. They had pretty much worn their spacecraft out.

By now, both astronauts had successfully gotten back into their spacesuits. Yes, I said "back into their spacesuits." You see, officially Jim Lovell and Frank Borman were the first Americans to fly in their underwear when doing 14 days aboard Gemini VII, but the closely guarded secret is that Cooper and Conrad did it first.

Prior to their eight day mission the two crewmen had reasoned that there was no way that they could endure eight days in those G4C suits, so, as Cooper stated in his official May, 1998 NASA Oral History: "… Pete and I tried to get approval to fly in just a regular flying suit on Gemini V, since we were not going to be doing EVA and we thought we had developed enough reliability in the pressurization system at that time we could be able to fly in the flight suit. But, NASA wouldn't approve it. So, the first thing we go into orbit, the first thing that happened is our helmets and suits came off and we were in our undies."

Thus, although Walter Cronkite proudly announced that the crew of GT-7 were to be credited with the first sneeze in space and the first to fly in their underwear, the truth is, they only had the first sneeze.

Once in orbit, there was really no way for NASA management to know that the crew was no longer in their suits and Cooper and Conrad knew it. So, in spite of management's insistence that they must remain in their suits the crew did what good pilots always do— they flew it the way it worked best and just kept quiet about it. NASA management never knew the difference.

Since their OAMS system was now totally useless, the crew of GT-5 were forced to activate their RCS thrusters a full orbit early in order to ensure that they could be in the proper attitude for retro-fire. Currently they were on Ring A and were intending to use it for the remainder of orbital flight and through reentry while keeping Ring B in reserve.

As they passed over Texas, everyone not working the flight in Mission Control ran outside and looking up, they saw Gemini V zooming over looking like a tiny bright point of light.

"They want me to tell you," CAPCOM Jim McDivitt reported to the crew, "it looks like you're moving fast."

"Yes," Cooper replied, "we're really whistling."

"Can you tell we're BEF?" Conrad quipped.

"Looks like you're about three degrees off in yaw." McDivitt joked.

"No," Cooper bantered back, "that's wrong."

"Must be the sun angle." McDivitt answered.

Dr. Berry, the astronauts' Flight Surgeon, was continually concerned that the crew may be so tired that they would not be alert enough to handle the complexities of the reentry. On the previous orbit, over the Bermuda station, he had asked the crew how they felt about taking any "aid" for fatigue. Of course by this time the crew had truly had their fill of doctors and medical experiments, and Cooper replied with a wise crack,

"Make a house call."

That did nothing to negate the concerns of the flight surgeon and on the current pass over Texas, Dr. Berry again called up to the crew,

"Gemini 5 this is Surgeon, I want to check again that we're in agreement that we will not use Item B, is that affirm?"

Although my research has yet to fully prove it, we can guess that "Item B" was a Dextroamphetamine tablet, commonly used in the military as a "Go-Pill" to enhance alertness in times of high fatigue. Cooper, in fact, was ordered to take one prior to his manual reentry of FAITH 7. He nearly landed on the carrier, so we can assume it worked.

"Tell them we took one for the road," Cooper replied to Dr. Berry.

Ironically, there sat Gordo Cooper in the exact same circumstances he had experienced on his long duration Mercury mission— his spacecraft systems had nearly completely failed and it was time to nurse the depleted spacecraft back to Earth. This time, however, he had something that his Mercury FAITH

7 vehicle did not have— a working computer.

For eight days Cooper and Conrad had baby sat their vehicle's computer, pampering it at every data entry and crossing their fingers as it was turned on and off to conserve power. All of that in the hope they would not experience the same sort of malfunction that had happened to GT-4.

As GT-5 swung around at Carnarvon, Australia Mission Control transmitted up their final numbers which were received by the computer and then double checked.

Over Hawaii, Conrad jettisoned the adapter by hitting the SEP ADPT button at 7 days, 22 hours, 26 minutes and 51 seconds mission elapsed time. Then, exactly 49 seconds later the crew got retrofire and GT-5 was on the fast track back to mother Earth. The spacecraft entered blackout at 07:22:43:40.

As soon as the vehicle began the reentry their Flight Director Indicators or "FDIs" were showing unexpected conflicting information. Cooper's FDI Downrange needle was full-scale up but Conrad's showed normal rates. Cooper rolled into the heads-up attitude in order to bring the needle back on target, but they were going to be way short on their landing. Something was amiss with their computer and in his NASA Oral History, Cooper explained it in fun terms,

"Well, when we started our reentry, we realized very shortly that it was taking us in at way too steep a dive-in angle. Since the Gemini spacecraft had an offset CG; rolling one way would give you uplift and rolling the other way would give you downward lift, so you had a fair amount of maneuvering with the spacecraft. We realized that it was calling for too much downlift, and we were going to come in way short. So we went to full uplift and recovered as much as we could, but we still were 100 miles off. It turned out the ground, in all these intricate calculations, someone had calculated that the Earth rotates 360.00 deg per rotation when, in fact, it rotates 359.999. When you add that all up for 8 days of data, it gave us the wrong reentry data."

Splashdown took place 80 miles short of the planned landing point and left the guys in the "trench" at Mission Control scratching their heads to figure out what had gone wrong. Their timing was perfect and the crew should have been right on target. After a bit of slide-rule swiping and some suspension of disbelief, they finally concluded they had simply miscalculated the rotation of the Earth.

Meanwhile, Cooper and Conrad were quickly helicoptered to the carrier LAKE CHAMPLAIN, which was the same carrier that had recovered Alan Shepard's FREEDOM 7 Mercury flight. As the astronauts were being given a near sadistic going over by the NASA doctors in the carrier's sick bay, the Gemini V spacecraft was hoisted aboard the ship. When technicians opened the hatches the smell nearly knocked them over! Indeed, as Pete Conrad would always describe it, he and Cooper had spent eight days in a garbage can.

MEMORIES OF OTHER FOLKS) Ross; USA: "I was born in 1958, but I started following our space program when I was about 4 with the launch of John Glenn. My parents (who, like most Americans didn't know any better at the time) took me outside to look up into the dark night sky to try seeing John

Glenn flying over. Of course, the orbital plane of the spacecraft never took it this far north, and I now know that he landed during the DAY, but it still made an impression on me that I never forgot.

Later during one of the Gemini missions, I was in school in the 3rd grade and our teacher brought a TV into the room for us to watch one of the landings. I don't recall which flight this was. But I remember watching the very crude animation of the reentry on one of the networks, and they showed some of the thrusters firing. One girl asked the teacher what those were, and I spoke up saying they were the pitch and yaw thrusters. The teacher just looked at me as if I were from another planet for knowing such a thing. (I'd learned that from reading a children's book on Project Mercury)....Such things a kid remembers."

GEMINI V
8 DAYS OR BUST- THE FIRST MISSION PATCH

In the months prior to the Gemini V mission I was real busy doing what most kids in our tiny Saginaw, Michigan suburb of Sheridan Park were doing— spending as much time as I could riding my bike in the street.

The summer of 1965 was an era when pop-culture was all about spy TV shows, corny beach party movies and hot cars. Everyone, it seemed, was into race cars and dragsters— even the astronauts were nearly all car-crazies. Chief among those car-nuts were Gordon Cooper and Pete Conrad.

I was a budding car-crazy too, and my Dad, who was a motor head himself, fed my interest. Stunned by my second grade teacher's claim that I could not read and motivated by the school board's reading specialist who told my folks that I could indeed read, on an advanced level, and all I needed was material far beyond "Dick and Jane," Dad bought me issues of car-oriented comic books such as "CARtoons" et. al. I lived in those until I got my first issues of spaceflight books from the Science Service.

Additionally, 1965 was the summer of the Stingray Bicycle. These little bikes had a banana seat, truss rods, butterfly handlebars and a racing slick rear tire.

It seemed as if every kid in Sheridan Park had one… except me. I had a "Leave it to Beaver" style little black bicycle with a mock gas tank on the crossbar, fenders, plain handlebars and book rack in the rear. My folks didn't have the money to buy me a new bike and mine was "perfectly good."

Sigh as I might, it made no difference to my Mom, but my Dad figured it out. Yep, my Dad, who was a master at adapting and improvising, found that he could go to the local discount store and get a "conversion kit" for far less than the cost of a new Stingray.

On my birthday in the spring of 1965 Dad came home from one of his two jobs and in his hands he had it all— the truss rods, the butterfly handlebars, a banana seat and handlebar grips made of translucent plastic with silver metal flakes in them and most of all a racing slick rear tire!

In less than two hours my Leave it to Beaver bike was transformed into the hottest Stingray in Sheridan Park. Dad nabbed a joker from a deck of playing cards and a clothes pin from Mom's basket, clipped it into the spokes and I burned up Lexington Drive— the happiest eight-year-old on the planet.

Pete Conrad, Jim Lovell, Wally Schirra, Alan Shepard, Gus Grissom, Gordon Cooper and all the rest of the astronaut car crazies would have been proud, because as they trained for Gemini, I was doing what they loved to do, racing around the streets at high speeds. Okay, let's just say high speeds relative to my scrawny asthmatic eight-year-old self, but I did have a Stingray bike.

Meanwhile, as I put down little patches of rubber on

my street, completely outside of my scope of reference the prime crew for GT-5 was in their final weeks of training for their eight-day mission.

Although Cooper was a superstar astronaut from the Original Seven Mercury astronauts and Conrad was a rookie from the second astronaut group or, "The New Nine," both men had gone to survival training in the South American jungle together shortly after Cooper's FAITH 7 mission.

Since they had been paired together in the jungle for a few days and made to "survive" they knew each other and made a good team. One day while working on GT-5 they began kicking around the idea of a mission patch. Every military pilot group had its own patch and every special military mission had a patch, so why should spaceflight be any different?

With that in mind, the two astronauts commissioned a patch to be sewn up depicting an old Conestoga wagon similar to one that Conrad's father-in-law had made a wooden model of; the crew felt that it symbolized a pioneering spirit. The words "8 DAYS OR BUST" were emblazoned across the wagon's cover and the names of both crewmen on the outer edges.

Of course no one in NASA management had approved any sort of mission patch and the naming of spacecraft had been nixed with MOLLY BROWN so Cooper and Conrad were treading on dangerous ground. Hugely in their favor was the fact that in that era of spaceflight, astronauts were powerful celebrities while NASA managers were largely unknown to the American taxpayer. So the two GT-5 astronauts decided to have the patches made and just come out wearing them on launch day.

Just two nights before their scheduled launch the crew was invited to fly from the Cape to Houston in order to have dinner with MSC Director Bob Gilruth and NASA Administrator Jim Webb. The dinner was set to take place at Gilruth's home. The crew decided at that time to, as Cooper put it, "...confess our sins.." and they told their two bosses about the patch.

Webb blew a gasket! The mission patch was a clear deviation of his efforts to depersonalize the Gemini missions. A big argument erupted between the powerful Webb and the hot-tempered Conrad and it took both Gilruth and Cooper to separate the two who were a few cuss words away from an actual fist-fight!

(See Image 8, page 160.) Finally, Webb simmered down and Cooper explained to him that a mission patch was not just for the crew, but was also for the hundreds of people who worked the mission and supported the crew. It was theirs to wear and keep, to show they had worked hard to make Gemini V a success. It was hard for Webb to argue with that logic and he asked to see the patch itself. Unfortunately, the two astronauts had not brought one with them. The administrator ordered that one be flown to Washington DC the next day for him to see and he would then make a decision.

His decision was that the crew could have a mission patch on one condition— they had to cover up the "8 DAYS OR BUST." Webb's reasoning was sound- if anything should happen to cause the mission to be terminated prior to the eight day mark, the news media would immediately point to the patch as evidence that GT-5 was a "bust."

In haste, the crew had some fabric stitched over

the patch to cover the unapproved wording and the United States manned space program had its first mission patch. Although many decades later there can be found "mission patches" for all of the Mercury missions and the first two manned Gemini missions, these are simply whimsical phonies dreamed up long after the missions were flown. Some were made up as late as the 1980s. The first genuine NASA Mission patch was that of Gemini V.

GEMINI V
ONE SYSTEM AFTER ANOTHER

Launch for GT-5 was scheduled for the morning of August 19, 1965. Anyone who has lived in central Florida can testify to the fact that summer in the sunshine state equates to thunderstorms. However, these intense storms are normally predictable and start in the late afternoon. They, however, can also pop-up at any time and tend to be isolated to a given area.

Northern born tourists can find themselves driving suddenly in and then out of the most frightening thunderstorm they have ever seen while local residents pay little attention.

So it was on the scheduled day of the Gemini V launch that a thunderstorm grew above Launch Complex 19, but the launch crew decided to proceed with their activity anyway. The odds were the storm would expend itself and fade away before launch time. Cooper and Conrad were loaded aboard the spacecraft and the erector was lowered. But, the storm cell intensified and a bolt of lightning struck a power station on the pad and the resulting surge caused the spacecraft's computer to get odd. Thus the launch was scrubbed for the day.

Slipping the launch led to a Saturday morning

departure for GT-5. My normal summer time for a Saturday morning consisted of a diet of very up-scale TV watching such as Alvin and the Chipmunks, followed by "Tennessee Tuxedo" and "Quick-Draw McGraw" on UHF channel 57, then a fast change to channel 5 for my favorite show, "Fireball XL5."

By that time my Mom normally gave me the old, "It's a nice day out, get yer' butt outside and play!" line as she threw me out of the house. She used that line on summer mornings even when it was raining. In Michigan you have to take advantage of every single day of summer— no matter what the weather happens to be.

On Saturday, August 21, 1965 my cartoons were all pre-empted for TV coverage of the GT-5 launch. I watched it until it was gone and then Mom booted me outdoors so I could get on my Stingray bike and be gone too.

What none of us on the outside of NASA and what most people within NASA, other than Cooper and Conrad knew at that moment was that GT-5's troubles had already begun.

Just about 02:06 into the boost, the Titan II's first stage started to pogo oscillate. The two astronauts began to get a rough ride as the pogo reached a peak of +0.38Gs and lasted for 13 seconds until staging took place. NASA had set the maximum limit for pogo as +0.25Gs, but Cooper and Conrad rode it out— there was no way that these two were going to "bust" their eight day mission on just some rough booster activity. On the air-to-ground communications recordings you can actually hear Cooper's voice being shaken by the pogo. The second stage, however, was as smooth as a cheetah and sped the crew into a nominal orbit.

Three days later an investigation of launch records showed that the oxidizer stand pipes in the first stage had inadvertently been pressurized with only 10 percent of the required amount of nitrogen. It was concluded that this was a procedural oversight. The problem was rectified and no other Gemini Titan II boosters suffered from such a high pogo level during the remainder of the Gemini program.

Through their first orbit the crew focused on getting their spacecraft in order. They found there were several screws floating around in the cockpit and Cooper reported at least 10 to 15 washers.

As the GT-5 passed over the Guaymas, Mexico station the crew got involved in transmitting Conrad's blood pressure and shortly after that they got a surprise, unauthorized radio call— from their back-up crew.

Astronauts Neil Armstrong and Elliot See were flying over Texas in their T-38s, NASA 903 and 902 when they dialed in the GT-5 UHF frequency and put out a call to Cooper and Conrad! Although both sides of the conversation were not recorded and the entire exchange itself is not on the archived audio, the side from the spacecraft is in the transcripts. Thus it is up to you to fill in the blanks,

Conrad: "Hello NASA 903, 903. Gemini V here. We read you loud and clear. Do you read us?"

NASA 903 and 902: ......

Conrad: "Well, we're right on the flight plan. We got all the gear unstowed and we're lookin' real good... fuel... right down the line."

NASA 903 and 902: ......

Conrad: "You'll have to climb pretty high."

NASA 903 and 902: …..

Conrad: "I'll bet our flight's more fun than yours."

Cooper: "Yes."

NASA 903 and 902: …..

Conrad: "I just changed my mind; spaceflight's better than flying."

Six minutes later,

Cooper: "903 Gemini V, see you guys in eight days."

Later Conrad asked CAPCOM Jim McDivitt if he had heard their conversation with "the back-up… NASA 903 and 902…" McDivitt replied that he had and it was "real good."

As far as my research shows, that was the first radio conversation between a crew in orbit and aircraft in flight outside of the approved Mission Control communications network.

One of the first items on the schedule for the GT-5 crew was to eject the Radar Evaluation Pod (REP). They successfully got it deployed and were busy evaluating their results when at 02:23:19 mission elapsed time Conrad suddenly noticed that the oxygen pressure on one of the fuel cells had dropped. With the automatic heaters on in the system, the pressure should not be dropping. Just an orbit and a half into their eight-day mission the crew had encountered a problem that could end the mission in less than one day.

By the next orbit, it appeared as if the fuel cells in GT-5 were rapidly dying and Mission Control ordered the crew to go into a power down mode. In fact so extensive was the power down that ground controllers were transmitting up instructions along with the statement, "no need to respond." They felt that just the act of keying the microphone in the spacecraft and transmitting a simple "Roger" required too great a use of power.

What happened next depends on which one of two different accounts of the events you want to consider as correct, NASA's official account or Cooper's Oral History account. Or, perhaps you may want to put the two together.

NASA's official story is that Flight Director Kraft immediately called upon McDonnell to set up a test where it could be discovered if the fuel cells would run on very low pressure. As he waited for the contingency reentry areas to be plotted, McDonnell ran the tests and came back with the results that the fuel cells would indeed run on low pressure.

Cooper's account is quite different,

"… This is the story a lot of people don't know: when Pete and I were going through the altitude chamber with Gemini V, we had to go through and do these various tasks; and the tests finished on Friday; the spacecraft was due to be shipped Saturday morning to the Cape from St. Louis in order to stay on the time schedule.

"But one of the things we wanted to do was, we wanted to run some tests in the altitude chamber by decreasing both oxygen and hydrogen pressure, drastically, to see if the fuel cell would continue to run at altitude. NASA said, 'No, we can't afford the time if we do it over this week, we can't afford to delay until next week to do it. And if we do it over the weekend, it would cost us triple time of overtime, so we're not going to do it.' So, Pete and I went to Jim McDonnell, head of McDonnell aircraft, told him the story on it, and he said "I'll pay for it. Let's do it." So we spent the weekend in the altitude chamber at his cost, doing the test; and if we had not done that test we would have reentered an orbit later. But because we had done the test, we knew the fuel cell would run at low pressures, so we decided to try it for a while and baby it, and see if we could get the pressure back up. And a few orbits later, they began to get indications that we were going to get the fuel cell back online."

In reading both accounts it is pretty clear that when Kraft asked for the testing and results and then got them in less than a full orbit, what he got was the results of Cooper and Conrad's weekend in the altitude chamber with their spacecraft— all of which was done at McDonnell's expense.

So, it would appear that both accounts blend together to give us a picture of what actually happened. Once the pressure stabilized at a drastic low of just 71 pounds, the fuel cells ran just fine and the crew began the process of monitoring and nursing their power source for the next eight days.

For the next two days, the crew went around the Earth conserving power, yet attempting to also get some experiments and photography done. They also managed to do some practice rendezvous with imaginary points in space done as well. For the most part photography had taken up their time.

Houston had planned for the astronauts to sleep in shifts, but in practice that just did not work. When one crewmember was trying to sleep even the most lightweight activity by the other crewman would disturb the sleep.

Once on day three while Conrad was simply speaking softly on the HF radio as a part of an experiment, it kept Cooper awake. Finally at the start of day four Conrad told the ground that both he and Cooper were so short of sleep that they both just dozed off at the same time. The flight surgeon decided on the spot that for the rest of the mission the ground would just go silent during the crew's normal Houston night time.

(MEMORIES OF OTHER FOLKS) Mary Ann; Houston, Texas: "When I was little my Daddy worked at the space center. He always worked the night shift and one evening before he went to work he told me that the news said that Gemini V could be seen tomorrow morning. He asked if when he got home, did he want me to get woke up so we could watch it go overhead? I said 'yes' and was so excited that I could hardly sleep.

I thought that we would see the full sized Gemini go zooming by just above the tree tops. When Daddy got home he came in and woke me up and we went out and stood in the back yard looking up. It was still dark outside and I wondered how we would see it in the dark. Then Daddy got very excited and pointed into the sky saying, 'There she is!' Looking up all I saw at first were stars, then I saw a bright one moving very fast. Puzzled I asked, 'Is that it? The moving star?' Daddy said it was. I'd expected a big spacecraft with two astronauts waving at us out the windows, but it was just a star. Later I learned that they were really far up and the sun shined on them making them look like a star."

Both crewmen seemed to be enjoying the "visual acuity" experiments the most. From their seats they were supposed to simply look down and see stuff.

There were two rocket sled runs at White Sands, New Mexico that had to be observed, the aircraft carrier LAKE CHAMPLAIN steaming in big circles and two Minute Man missile shots from Vandenberg Air Force Base that the crew excitedly saw and photographed. For a while it looked like the fun would never end, but then came the collapse of another critical system.

As the spacecraft was acquired by the Carnarvon station, Conrad came on the radio at 04:22:31:50 and asked if they were ready to copy "…a little problem." GT-5 had been out of communications range from the Canary Islands to Carnarvon, a distance of nearly half an orbit, when Conrad discovered that the left number 7 yaw thruster was simply not working. Although it sounded like a simple glitch at the time, it was the first domino to fall in the OAMS thruster system. After some routine switchology and circuit breaker cycling, both astronauts were soon scheduled to go into their sleep period while Houston worked the problem.

In an effort to conserve their remaining 12% of OAMS fuel the crew was given the order to shut down the OAMS system and go into drifting flight. The problem was that their fuel cells required venting and that little vent acted like a tiny thruster. This resulted in an end-over-end tumble of as much as 12 degrees per second. Yet while Cooper and Conrad sat there, almost totally powered down in what Wally Schirra called "chimp mode" slowly flipping end over end, Mission Control still wanted them to try and conduct photography. The big problem there was that any time they were passing over something interesting, they were pointing the wrong way.

To make matters worse, over the next 57 hours the

rest of their OAMS thrusters began to follow number 7 and die.

Things began to get a bit "testy" as ground sent up requests and the crew was in no mood to comply. At one point Mission Control asked Conrad for an oral temperature and he did not reply. Cooper said that his pilot was "busy" and the ground told Cooper to just stick the thermometer in Conrad's mouth. Cooper replied that they would just have to try it at another station.

Both astronauts had cramped legs, aching knees and were shivering cold, yet determined to make it to eight days. Ironically, there sat Cooper with his spacecraft systems dying as he tried to make the full length of his planned mission— just like his Mercury Faith 7 flight.

Meanwhile, Conrad was contemplating the worst decision he had made in the last decade or so. Before the mission he had considered bringing along a good book to read— but then decided against it.

GEMINI V
THIS IS GEMINI V ON THE WATER!

At 07:23:03:35 mission elapsed time Gordon Cooper gleefully radioed to Air Boss 1,

"This is Gemini V on the water!"

He and Pete Conrad had done it. They had set the world's record for long duration spaceflight. They had also compiled a huge laundry list of changes and things to avoid for the crew of Gemini VII who would soon attempt to double their record. Top of that list was trash management and somewhere on the list was bring a good book to read. (The crew of Gemini VII brought "Roughing It.")

Cooper and Conrad were dispatched on a world-wide friendship tour shortly after their flight. Then, upon their return to Houston, Conrad went immediately into training with Dick Gordon. Conrad was to be the back-up commander for Gemini VIII and Gordon was to be the back-up pilot. Later Conrad would serve as commander for Gemini XI, with Gordon as his pilot.

Meanwhile Cooper was relegated to back-up commander for Gemini XII and later back-up commander of Apollo 10, but he never did fly again after having fallen out of favor with Deke Slayton and much of NASA's upper management. It was felt that Cooper was focused more on hot cars and racing than he was on spaceflight training. Slayton later recalled that Cooper was simply, "...a hard sell to management," when it came to crew selection.

The fuel cell problem on GT-5 was solved by the addition of a single component and they went on to fly well on future missions. The issue with the OAMS thrusters turned out to be much easier to fix. When the fuel cell problem came up on GT-5, the emergency power-down included the heaters on the OAMS thrusters— which sat unheated for nearly four days. On the next endurance mission, GT 7, the heaters would simply be left on— no matter what.

GEMINI
CSQ, RKV, RTK… ECT

As an aid to air-ground communications and especially telemetry, tracking ships were used by NASA. In the case of the Gemini flights the vessels used were the US Naval Vessels ROSEKNOT and COSTAL SENTRY.

Additionally the Air Force Systems Command's vessel RANGE TRACKER, which had been on missile tracking duty along the Eastern Test Range since 1960 was pressed into Gemini service.

These vessels were stationed in areas where ground station coverage was not available. Good examples of those open areas would be the southern Pacific Ocean as well as off the coast of Red China. The Pacific posed a problem simply by its great expanse and Red China posed a problem because they were really grumpy and not really open to the idea of NASA placing a tracking station there. In fact they were not open to most ideas— so it was best to post a tracking ship a significant distance from their shore.

Naturally, all NASA tracking stations had to have a three-letter identifier assigned to them. This supposedly made it easier for teletype communications and radio communication. Ground stations, although officially assigned a three-letter identifier, such as

Hawaii's "HAW" and California's "CAL" were called by their full name when voice was being used yet the tracking ships were always called by their three letter moniker. Of course there was just one problem— none of the ships had three word names.

NASA being NASA came up with a solution so obtuse that to this day I firmly believe it was cooked up by a few Goddard engineers on a Friday night drinking at a local bar in Greenbelt. They took the COSTAL SENTRY and added the word "QUEBEC" to the moniker and its call sign became "CSQ." Then they took the ROSEKNOT and broke its name in half to give them "ROSE KNOT" and added the word "VICTOR" making her call sign "RKV." Finally they took the RANGE TRACKER and left her first name alone but broke up "TRACKER" and made it "TK" thus giving her the call sign "RTK."

All three of these ships served NASA until the end of the 1960s when they were taken out of commission.

RANGE TRACKER was removed from duty in 1969 and scrapped in 1970, COSTAL SENTRY was removed from service in 1968 and scrapped in 1969 although her registration paperwork was not converted until 1972. ROSE KNOT was decommissioned in 1969 and remained in lay-up until she was turned over to the United States Maritime Administration (MARAD) in 1977.

GEMINI VI
WALLY'S VIP GUEST

In preparation for their rendezvous and docking mission with an Agena target vehicle, the prime crew for Gemini VI, Command Pilot Wally Schirra and Pilot Tom Stafford spent more than 20 hours flying a contraption called the "Gemini translation and docking simulator." The device consisted of a full sized Gemini spacecraft reentry module and a full sized Agena mockup that were suspended in a web of steel girders and cables. Set up in 1963 at Building 1244, a huge hangar at the Langley Research Center, the simulator allowed the astronauts to sit in the spacecraft and fly a three dimensional approach to dock with the target vehicle.

Running along a track in the ceiling, the truss that supported the Gemini mock up used electric motors to travel backward or forward for a distance of 210 feet at speeds as great as 20 feet per second. It could also move the Gemini sideways a distance of 16 feet at four feet per second and up and down for a distance of 45 feet at a rate of 10 feet per second. Those high rates, however, were very rarely used. Normally the final approach would be a slow process using only tiny thruster inputs. As a result the simulations took a good deal of time. Every input from the pilot was fed into a fairly primitive electronic control system on the hangar floor and then transmitted by wire to the motors that made the machine go.

A "gimbal" that held the Gemini spacecraft mock-up itself allowed it to roll, pitch and yaw. The pitch and yaw rates were one radian per second and the roll rate was two radians per second. Thus the overall effect was considered to be pretty darned close what the astronauts would encounter in space. It cost NASA $320,000 in 1963 dollars to have the simulator constructed and considering that after Gemini it was adapted for use in Apollo, it was a pretty good investment. By 1985 the simulator had been sitting, unused, for about a decade in Building 1244 when it was declared a National Historic Landmark— a status that it retains as of this writing.

(See Image 9, page 161.) One day while at the simulator Wally Schirra was notified that he was about to receive a VIP guest. Hubert H. Humphrey, the sitting Vice President of the United States wanted to see the simulator in action and to also go for a ride in it. NASA officials of every stripe were there as were plenty of news media.

After being helped into the right hand seat of the Gemini and being strapped in for safety, Humphrey's hatch was closed and so was Schirra's. The Mercury veteran astronaut briefed the Vice President on how the simulator worked and what to expect and when Schirra was finished talking Humphry looked at him.

"You won't tell anyone if I take a little nap while you do this?" the Vice president asked.

Apparently Humphrey had been traveling quite a bit and he was exhausted. Schirra happily agreed to keep the secret and then went about flying a nice long docking profile while the Vice President snoozed. Meanwhile, the news media and NASA PAO took movies and shot photos as NASA executives proudly watched having no idea that Humphrey was napping.

When the simulation was done and the rig was reset Schirra woke the Vice President up and Humphrey, the pure politician, exited the Gemini giving praise to everyone involved— especially Schirra. No one but he and Humphrey knew that it had been a protracted nap.

"After that, I always liked Humphrey," Schirra would confess many years later.

GEMINI VI
THE SITUATION IS NOT A HAPPY ONE

On the cold morning of Monday, October 25th, 1965 I was seated at my desk inside Nelle Haley Elementary School's "Annex 3" as Mrs. Bechtol, my third grade teacher, attempted to hold my fleeting attention so that she could practice her craft upon me.

Opened in 1958, Nelle Haley had been quickly overwhelmed by the boom of residents rapidly moving into Sheridan Park. The subdivision itself had started construction shortly after World War II and by the mid 1960s only ¼ of the lots were still vacant and those would fill up before 1970.

Meanwhile a new subdivision had sprung up just outside the park and every new home meant at least two or three new students enrolling into the tiny school. In order to cope with the student boom the school board had acquired three small houses near the school and after ripping out the inner walls they monikered the make-shift classrooms as "Annex 1, 2 and 3" and I was unlucky enough to be assigned to one of them for the third grade.

By late October I was so bored I could scream and spent most of my time looking out the window— which was the one benefit of the Annexes, they each had one large picture window.

Any diversion from our standard third grade routine was a welcome change as far as I was concerned. Once Mrs. Bechtol got ticked at me and made me go "stand in the corner" while everyone else did math. I thought that was great, all of the other kids had to do school work while I got to just stand there and ponder the previous evening's episode of "Gilligan's Island."

Today's schedule of Annex 3 activities sounded even better. After taking the attendance the teacher announced that later this morning we would all be walking over to the main school so that we could watch a space launch on TV. Indeed another Gemini launch was going to rescue me from another catatonic morning of spelling, math and perhaps the worst thing of all— reading aloud. I was so filled with cheer that I might have even stopped doodling on my scribble paper.

Down at Cape Canaveral's Launch Complex 14 just 1,048.26 miles from my perpetually boring desk in Mrs. Bechtol's classroom, the final minutes in the countdown of an Atlas Agena launch vehicle were ticking away.

This Atlas D, vehicle 5301, was slated to place into orbit the Gemini Agena Target Vehicle, (GATV) 5002, for the crew of GT-6. Their mission was to accomplish the first ever rendezvous and docking of two spacecraft.

(See image 10, page 161.) Atlas 5301 had been on Pad 14 since August 16th and was now ready to fly with the GATV mounted atop. At 10:00 in the morning on October 25th the Atlas ignited, lifted off and burned for 04:44 which was just three seconds longer than predicted. From the shutdown of the

Atlas' sustainer stage the vehicles internal timer counted off 26 seconds and then the Agena itself was separated and went into a scheduled 40 second coast. What followed was about to wreck the whole morning for me and for the folks in Project Gemini as well.

After the Agena's 40 second coast period the target vehicle was programmed to ignite its Primary Propulsion System (PPS) engine and insert itself into the prescribed orbit. When that event was due to take place, suddenly almost all telemetry from the Agena simply stopped.

At first the Agena Controller in Houston reported "a momentary loss of signal." Upon a second look he saw the pressure in the PPS coming up and then the signal was abruptly lost. Ground tracking stations began reporting "no joy" when asked if they could see anything from the vehicle. The Bermuda tracking station continued to get an S-band radio carrier signal from the Agena and the Antigua station got a C-band signal which gave everyone some hope. But there really was no hope, the best guess was that when the PPS ignited there was a catastrophic explosion that simply blew the vehicle apart. The S-band and C-band transmitters were simply still connected to their power sources and tumbling among the bits and pieces of what was once the Agena. GATV 5002 was simply space junk.

NASA's Paul Haney in announcing the status reported,

"The situation here is not a very happy one..."

As the time came for the Agena to pass over the tracking stations at Australia, the story was the same— no telemetry. With that, Schirra and Stafford, who had been loaded aboard their Gemini spacecraft just under two hours earlier, were given the news that the GT-6 mission was officially scrubbed. The situation there was not a very happy one either.

A few minutes after word of the scrub was made public, the black telephone in Annex 3 rang and Mrs. Bechtol answered. She spoke, hung up, shuffled some papers and then informed the class that we would not be walking over to see the space launch because it had been cancelled. Well, that wrecked my whole morning. There I sat as Mrs. Bechtol ordered the whole class to get out our spelling books. The situation was not a very happy one.

Within a week it was concluded that the loss of GATV 5002 was caused by a "Hard Start" of the PPS engine. The PPS engine used hypergolic propellants, which combust on contact. Although more than 180 Agenas had flown prior to Project Gemini, all had led their start sequence by use of the oxidizer feed line. For the GATV Agenas, however, Lockheed had decided to lead with the fuel feed line injecting into the combustion chamber. That change led to the loss of GATV 5002 and reversing that change, it was thought, would eliminate the problem.

Shortly after the scrub was called, Schirra and Stafford returned to the crew quarters on Merritt Island. Their GT-6 spacecraft was the last Gemini to be equipped with battery power alone, all future flights were scheduled to run on fuel cells plus batteries. Thus, without an Agena, the crew had no mission. Schirra had been campaigning to get the first rendezvous since his Mercury flight— this was his baby and now it was gone. As the crew got aboard their T-38s and flew back to Houston, the situation during that flight was not a very happy one.

GEMINI VII AND VI-A
THE BURKE-YARDLEY CONCEPT

Even as Schirra and Stafford were flying back to Houston following the scrub of GT-6, plans were being hatched that would brighten their future.

While the bits and pieces of Gemini VI's Agena were reentering the atmosphere, McDonnell's Spacecraft Chief Walter Burke rounded up his deputy John Yardley at the Cape control center and began a quick discussion that involved using Gemini VII as a replacement target vehicle for Gemini VI.

Shortly after the discussion began the two men pulled Frank Borman, Gemini VII's command pilot aside and brought him into the plan. It was all based on a recent study for a quick turn-around of a Gemini Titan II vehicle and soon Raymond Hill, McDonnell's man at the Cape who was in charge of the hands-on movement of the hardware was brought into the loop. Hill was familiar with the plan and brought everyone else up to speed. Next Burke and Yardley took their idea to NASA's upper management and that was where the roadblocks went up.

Management felt that there were too many "impossibles" in the Burke-Yardley scheme and that the best plan was to simply stack the Gemini VII spacecraft atop the Gemini VI booster and fly that mission alone.

But then the engineers did the math and found that the Gemini VII spacecraft was too heavy to be boosted by the Gemini Launch Vehicle (GLV) 6 booster. (Each Gemini Titan II was constructed to loft a specific spacecraft.)

Only then did the Burke-Yardley plan begin to gain traction. Soon the roadblocks began to fall and eventually the plan found its way to NASA Administrator James Webb.

Unlike his underlings, Webb was all in favor of the plan; it was bold, it was forward thinking and although risky, it would make history. This was an era when NASA's administrator pointed his agency toward doing bold and historic things. He approved the plan and forwarded it to the President. LBJ also approved and just three days after their mission had been scrubbed, Schirra and Stafford had their mission back.

On October 28th, the same day the White House made the announcement that Gemini VII and Gemini VI would rendezvous in space, Gemini spacecraft 6 and the GLV 6 second stage were deerected at Launch Complex 19. Then, the following day, Thursday October 29th, the GLV 6 first stage was also deerected. Spacecraft 6 was stored in the Pyrotechnics Installation Building at the Merritt Island Launch Area while the booster was stored under guard at the Cape Canaveral Satellite Checkout Building where it was kept in a climate controlled environment. This bonded storage was done to maintain the integrity of the pre-launch check-out of the booster. Now that equipment would wait until after Gemini VII had been launched. The

Burke-Yardley plan was now in full motion.

(MEMORIES OF OTHER FOLKS); David, Fargo ND: "My memories of Gemini are very good because I was almost 6 years old when the Gemini 6/7 mission went on glued to the TV in glorious black and white. But my mother started a Gemini scrapbook for me and I have cherished it ever since and I have spent my hobby time backtracking and getting all the Gemini history I can and of course the rest of the space program. But this is how my interest in the space program started."

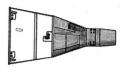

GEMINI VII AND VI-A
ABSOLUTELY A GUINEA PIG!

If anyone considered Gemini V's eight day marathon to be boring then Gemini VII would be double boring.

Command Pilot Frank Borman and pilot Jim Lovell were launched at two seconds past 2:30 on the afternoon of Saturday December 4th, 1965. Although I had just come inside from playing in the fresh snow and had just shed my wet boots, socks, pants, mittens, hat, jacket and shirt in order to thaw my benumbed toes over the heater vent, the final countdown to the launch got my attention.

Borman and Lovell, however, were far better dressed than I had been. They were sporting the new G5C lightweight spacesuits. These suits had no attached helmet, but rather sported a soft hood with a faceplate. Inside the hood each astronaut wore a helmet similar to the standard helmet worn by fighter pilots of that era.

The suits were supposed to have greater comfort in the cockpit of the Gemini spacecraft and were also supposed to be easily removed and quickly donned in the event of some sort of cabin depressurization; neither was true. In fact, the suits were hot and uncomfortable in flight and once they were removed there was no way to get them back on quickly— it took the better part of an orbit.

Still, Borman and Lovell rocketed into space seeking to set the world's record for spaceflight and prove that humans could live in space for extended periods.

Although immediately after separation from their Titan II's second stage the crew did manage to station keep with the spent booster for a while, no matter how anyone looks at it, the flight of GT-7 was purely a medical mission. When asked years later did you feel like you were an experimental rat up there...? Lovell replied,

"A guinea pig. Absolutely a guinea pig!"

They were, however, much more prepared to avoid the garbage can syndrome that Cooper and Conrad had experienced. Both crewmen had spent countless hours in the Gemini trainer figuring out how to take every single bit of waste paper or discarded container and painstakingly fold it as small as possible and then stow it. They spent as much time doing that as they did practicing some of the flight maneuvers. Thus they started their mission with a very clear view of the problems ahead— and that would pay off.

GEMINI VII AND VI-A
FAST STACKING A TITAN

About as fast as the last little fires on Launch Complex 19 were quenched and the remnants of the toxic propellant residue was drained into the skimming basin, the booster for Gemini VI was cracked out of bonded storage and trucked to the pad.

Stacking a Titan II booster has always been sort of an overlooked process. The way in which it was done, however, is quite interesting.

The first stage was rolled to the pad on a eight-wheel trailer pulled by a "Six-By" military truck. The rig was then backed up the pad access ramp to the erector, which was resting on the pad in the lowered position. Two steel temporary ramps were in place to allow the wheels of the trailer to be rolled onto the erector.

The wheels of the trailer were boat-trailer sized and placed in dual sets so, in order to keep the stage properly aligned four square "U" shaped wheel troughs were constructed as a part of the erector and ran the full length of the structure matching the wheel-base of the trailer.

Once all of the wheels were in the troughs, the truck was un-hitched and the stage and trailer were winched into a location that would be slightly above the launch pedestal support ring. The stage itself was transported with a support ring fastened around it. A web-like structure was fastened to the top of the stage. These were secured so that the stage would remain in place while the erector tilted it into the vertical position.

The trailer was then rolled out from under the stage and the two temporary wheel ramps were removed. Two lifting arms, which were huge "I" beams that extended below the pad, were then actuated and hydraulically pushed the erector into the vertical position. Keep in mind that the empty weight of the Titan II first stage was only about 9,583 pounds and the empty weight of the second stage was only about 5,099 pounds, so movement and lifting of these components was not an enormous task. (Fully fueled, after being stacked, the booster stack weighed more than 316,000 pounds).

Once the booster had been raised to a vertical position, the erector's crane, by way of the forward web-support, lowered the stage onto the launch ring were it was bolted to the pad. The support rings were then all removed and the erector was then lowered to receive the second stage. That stage was moved upon a trailer which also had eight wheels in four dual sets. The stacking process was exactly the same other than the fact the stage was bolted to the top of the first stage rather than to the pad itself.

(See Image 11, page 162.) The erector's crane was then cantilevered out to receive the Gemini spacecraft. It was lifted from its trailer to the deck of the launch pad by way of a small gantry crane and once placed upon the deck it was lifted to the white room at the top of the booster by the erector's crane and moved horizontally into position where it could be lowered atop the second stage.

In the hours of darkness between Saturday and Sunday the GT-6 launch vehicle was fully stacked and by mid-day on Sunday the Spacecraft was in place.

Meanwhile technicians were busy repairing the pad damage done by the departure of GT-7. Major damage had been done to the main electrical conduit and the line that fed hydrogen to the spacecraft's fuel cells was blown away. The lucky part was that the electrical lines were easily patched around and the GT-6 spacecraft was battery powered so it had no fuel cells to feed hydrogen to and thus that line was not required.

Now the big question was whether or not the systems and spacecraft logic that had been functioning so well back on October 25th would still check out as ready to fly.

While the check-out of booster USAF serial number 62-12561, the Gemini VI Titan II, moved forward, Schirra and Stafford's mission was given a slightly new moniker— Gemini VI-A.

Although the crews for both GT-7 and GT-6A were delighted at the two manned spacecraft rendezvous plan, NASA planners along with Burke and Yardley had some elements in mind that the crews immediately nixed.

The first one was actually a back of an envelope sketch by Burke that he did while standing in the control center at the Cape back on October 25th. The scheme consisted of an inflatable docking collar that would be attached to spacecraft 7 in the aft center of its adapter module. Borman saw that and immediately told the two engineers that there was no way that he was going to have 7,833 pounds of Gemini spacecraft sticking its nose into the adapter where his life support, fuel and power were located. At that point in time no one even knew if they could actually fly together in a stable formation so attempting to dock with GT-7's adapter was just nuts. Borman dug his heels in on that one and the scheme was dropped.

A second scheme popped up a few weeks later when it was suggested that a double EVA should be in the mission plan. The thought was that while the two spacecraft were in close station-keeping Stafford and Lovell should go EVA and swap spacecraft. That one gained some traction until it was revealed that Stafford, being one of the tallest of the astronauts, was having a real hard time getting in and out of the hatch while practicing in the zero-G training aircraft. Once there was a hint of trouble in the activity, the whole scheme was dropped. GT-6A would fly up, rendezvous and then come back down— that was the mission... period.

Author's composite

1 **Lt. Col. Albert shows America the inside of the blockhouse at Complex 19.**

NASA photo

NASA Public Affairs Officer Paul Haney the voice of Gemini launch control. 2

3   **Young (left) and Grissom (right) at Pad 19.**

Wally Schirra, "Ready to go" as GT-3 back-up.   4

Dan Goodsell photo

In its time Gemini became such a part the culture
that it showed up in toys and even food containers.   5

McDivitt and White leav the suit-up trailer at
Complex 16 to take a ride on the Barton FREEway.

6   NASA photo

Ed White, the free world's first spacewalker. 7
NASA photo

NASA photo

**NASA's first crew mission patch,
that almost caused a fist flight.** 8

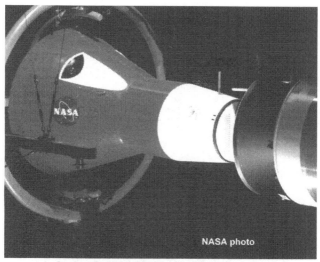

NASA photo

9 **Langley docking simulator.**

NASA image

**A fairly early concept drawing of a
Gemini and an Agena target vehicle.** 10

GT-1 second stage being stacked by use of the erector at Pad 19. 11

NASA photo

Gemini VII as seen from Gemini VI. 12

NASA Photo

Gemini VII (left) and Gemini VI come ashore.
Background, upper right "The Big Umbrella." 13

**Neil Armstrong and Dave Scott sail safely
into port aboard the USS LEONARD F. MASON.** 14

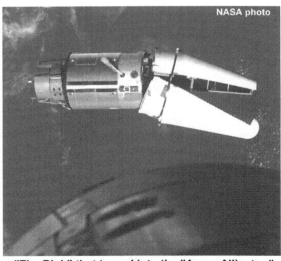

NASA photo

"The Blob" that turned into the "Angry Alligator." 15

Author's composite

**Simulated astronaut Miles McClure
breaks the door on CBS' Gemini mock-up.**   16

NASA photo

The Astronaut Maneuvering Unit (AMU)   17

Gemini X crew, Mike Collins (left) John Young (right). In the background is LC-19 and to the left is the Barton FREEway. 18

19

Dick Gordon preparing to jettison equipment aboard GT-11. When I was a kid this onboard image fascinated me- I studied for hours.

Buzz Aldrin training in the McDonogh pool. 20

I didn't win. 21

GEMINI VII AND VI-A
OUR MOST EXCITING DAY IN SPACE

At 9:00 Sheridan Park time on Sunday morning, December 12th Walter Cronkite opened CBS news' broadcast of the launch of Gemini VI(A) by cheerfully stating,

"Good morning everyone, this ought to be our most exciting day in space…"

He had no idea just how true those words would become in less than 54 minutes.

Over his left shoulder was a TV screen and on it was a live picture of the GT-6A launch vehicle and spacecraft poised for ignition and in his voice was a note of gleeful excitement. In that spacecraft were Schirra, Stafford and about 301,771 pounds of highly toxic, corrosive propellants that would burn on contact. There was, however, one tiny plastic item that did not belong there— and no one knew it was aboard.

Sometime in the spring of 1965 a technician at the Martin Baltimore plant had removed one of the engine's gas generators on vehicle 62-12561 for cleaning. As the check valve for the oxidizer inlet was separated, he placed a small plastic dust cover in the gas generator port in order to keep dust and dirt out. When the gas generator was later reinstalled, probably by different personnel, the little dust cap was overlooked and once the entire assembly was put back together there was no way to see the cap and there were no preflight tests that would indicate that it was there. The result would be that at about the same moment that the bolts blew and the Titan II began to rise, the thrust would have decayed to the point where the engine would shut down and the whole stack would come crashing back to Earth in a huge fireball garnished with a toxic orange cloud. Now, as Cronkite announced the this and that's of the mission the clock counted down toward that disaster.

Schirra's hands were on the "D" ring that when pulled would eject both he and Stafford simultaneously from the spacecraft. Once the erector had been lowered, ejection was their only way to escape. Stafford also had a "D" ring, but he had concluded long ago that he would leave the ejection up to his commander.

Most Gemini astronauts had little faith in the ejection system, feeling that their oxygen soaked suits would light up like roman candles with the firing of the ejection rockets. Stafford referred to it as committing suicide to keep from being killed.

In the spacecraft the astronauts listened as the Test Conductor, "TC" Frank Carey counted off the final events. At T-minus 02:00 TC called that the big first stage pre-valves were opening. Now the oxidizer from the stage's upper tank flowed down as far as the engines and waited for the ignition signal. At T-minus 01:30 the TC called that the Titan II was going onto internal power. Next at T-1:20 TC notified the crew that the engine gimbal test was taking place. In the cockpit the crew did not really need a notification. The action of the engines gimbaling through a pre-programmed pattern of movement was both felt and heard in the cabin. In fact it was a loud and obnoxious grind and groaning that shook the vehicle and really got their attention. Every crew commented on it during their post-flight debriefing.

Finally at T-00:35 in the count the second stage's pre-valves were reported opening by TC. The crew heard the big valves open and also heard the "glug, glug, glug" of the oxidizer flowing down the pipe. The big bird was ready to light her engines.

From the Complex 19 blockhouse astronaut Alan Bean was serving as "Stony," the blockhouse CAPCOM, and it was his job to count down the final ten seconds to the crew and call ignition. Bean counted down exactly as he was trained and when he saw the signal he called out, "Ignition."

With an almost primordial scream like a giant monster coming to life the Titan II's turbo pumps spooled up to speed and fed the hypergolic fuel and oxidizer into the combustion chamber. A great orange cloud of toxic smoke burst from the pad's flame bucket as the near explosive burning of the propellants blasted from the twin Aerojet engines and the entire engine mount shook. At nearly that same instant the hand of fate reached out and saved the mission of GT-6A as a small cannon plug at the base of the booster shook loose and dropped out.

That plug was meant to pull out at liftoff 3.7 seconds after ignition, but it dropped out nearly immediately. To the malfunction detection system, that error was clear and the signal was sent at the speed of light to shutdown the entire launch system. Instantly the engines were shut down, the thrust tapered off rapidly and soon all that could be heard was the low shrinking howls of the turbo pumps spooling down as if the monster was now falling asleep.

In the cockpit, Schirra's clock had been started by the dropping out of the tail plug. This was an indication of liftoff, but if a liftoff of less than an inch had indeed taken place and then the engines had shutdown, the Titan II was no longer bolted to the pad and was about to topple over and explode— it was an ejection situation. Yet Schirra saw that things did not add up.

The actual liftoff was supposed to happen 3.7 seconds after ignition, not immediately at ignition. Additionally, none of his attitude needles were moving and he had no feeling in the seat of his pants that the vehicle was moving. Finally, the sound of the thrust had rapidly come up and then vanished just as rapidly.

In less than two seconds Schirra, the experienced test pilot and Mercury astronaut correctly calculated that the were still bolted to the pad. Glancing at his fuel pressure gages he saw that they were beginning to drop- he and Stafford were indeed still bolted to the pad.

"Shutdown Gemini Six!" the TC rapidly called to the crew, "Program reset!"

Now Schirra knew for sure he had not lifted off. His major concern now was the tank pressures. If they stopped venting, the crew may have to eject anyway.

Across the free world, hearts stopped as everyone waited for the worst and for the crew to eject. Instead, Schirra and Stafford did exactly the right thing— which was to sit tight and do nothing.

In less than three minutes the Test Conductor sent word up to the crew that the indication was that of a tail plug dropping out. Without doubt the technician who had installed that plug was probably feeling pretty bad for the next 24 hours— until it came to light that he had unwittingly saved the entire mission.

Oddly, the mission rules said that after a situation such as this, if the crew was to be removed by way of the erector it had to be up and in place within 15 minutes. Schirra thought that was dumb and when the ground told him it would take 20 minutes he happily agreed.

As that 20 minutes turned into 99 minutes he advised that they should read the mission rules again. No one considered just how long it would take to fully safe the booster.

Again, oddly, there was plenty of experience with Titan II shutdowns. Test vehicle N-4 had a shutdown on Launch Pad 16 back on June 27th, 1962 when it suffered a "severe start transient." That issue actually blew the thrust chamber of one of the engines off of the mount and shot out through the flame bucket. It landed "…several hundred feet…" away from the pad. N-4 was successfully launched on July 25th of that same year.

Also, GT-2 had a shutdown on December 9th, 1964, just a year and four days prior to GT-6A's shutdown. In that situation the ignition took place and in the 3.2 seconds prior to the programmed blowing of the hold-down bolts and liftoff, the booster sensed a pressure drop in its primary hydraulic control system and instantly switched over to the back-up system. Since the engines were running, the malfunction detection system sensed that switch-over and it could not allow the vehicle to launch on just the back-up system, so it sent the shutdown command.

In both cases, the malfunction detection system did exactly what it was supposed to do and it saved the vehicle to fly another day. The fix required re-machining a component and the new part did not arrive at the Cape until January 9th and NASA had to slip the GT-2 launch to January 19th.

Immediately following GT-6A's shutdown, the launch vehicle sat there with two astronauts aboard plus its toxic propellants— some of which were currently venting into the surrounding atmosphere and to make the matter worse, about 1,830 pounds of that propellant had already been combusted into the pad's flame bucket. The entire scene was horribly toxic and highly dangerous, but the ground crews now had to get the crew evacuated as quickly and safely as possible.

It also had to be kept in mind that the booster and spacecraft were both bristling with live pyrotechnics. Scores of squibs, explosive bolts and linear shaped charges were everywhere. There was a slide wire that extended from the service tower to a safe area at the pad perimeter, but the crew had no way to get to it without the erector raised and in position. If they opened their hatches now, there was nothing but empty space. Getting to the slide wire required stepping into the erector's whiteroom and then dashing to the doors opposite the elevator, go down a flight of stairs and across a short walkway to the service tower. Thus, the first problem to be solved was raising the erector.

Pad technicians clad in Self-Contained Atmospheric Protective Ensemble (or SCAPE) suits were at the pad within minutes of the word that the vehicle was stable. They first had to replace the steel grid decking around the engines and then had to go about ensuring that the atmosphere was indeed safe for the extraction crew to move in from their safe positions. That crew would then raise the erector. This whole process all took an agonizingly long time.

Schirra and Stafford were left there "…just breathing," as Schirra reported to the TC. Their problem was that their spacesuits were at their worst when reclined there in 1G.

Schirra, contrary to what some other authors may say, was wearing the G3C suit, of the same model that Grissom and Young had worn and that suit had a single zipper up the back and along the spine. So after being in that seat aboard GT-6A for nearly three hours, he was quite uncomfortable.

Stafford, however, was wearing the newer G4C suit that had a double zipper up the back. By all rights he should have been in double agony, but he had thought ahead and ordered a long strip of wide foam rubber to run the length of his spine. That was inserted while he was suiting up and now it paid off. Still, by the end of their entrapment in the spacecraft, Stafford had to resort to grabbing the hatch closure loop and occasionally pulling himself off the seat in order to tolerate the discomfort.

After the crew had been in the spacecraft for a total of three hours and 22 minutes the erector finally started to raise. It took six minutes and four seconds for the erector to reach the vertical position and come to a stop; Schirra was livid.

Just 30 seconds after the erector came to a stop a team of three engineers dashed to the base of the structure and headed up the stairs. The elevator was locked at the spacecraft level to afford the astronauts an alternative exit once the erector was in place. Therefore the only way for the extraction crew to get up to the whiteroom level was a climb up the tower by way of 80 feet of caged rung ladder and then a series of stairs to the spacecraft level in the white room. Adrenalin must have been really pumping as the first member of the extraction crew made it to the spacecraft level in just two and one half minutes!

Quickly Schirra and Stafford were helped out of the spacecraft. Together they took the waiting elevator down to the pad level and were met by the emergency crew. This was an actual pad abort and the procedures called for the astronauts to be evacuated to the safe area near the blockhouse by way of the two armored vehicles that were waiting for them on the pad. So, although both astronauts were in good condition and by the time they had finished their elevator ride and Schirra's temper had cooled down, they were directed away from the waiting astronaut van and into the their assigned armored vehicle.

Once at the blockhouse safe area they were finally allowed to get aboard the astronaut van and head off to get out of their suits. The entire evacuation was shown on live TV as the Cape's remote cameras followed the process.

In watching the events it is a sure bet that many out there in TV land considered that Cronkite's opening statement was far too correct. This was indeed our most exciting day in space— but not for the reasons everyone had expected.

(MEMORIES OF OTHER FOLKS) Jeffrey, Melrose, MA: "I have some faint memories of Mercury, but I was too young. In 1965 I was 6 years old and the bug really bit. A few random memories...

I wrote to Chris Kraft with my enthusiasm, asking for any details about Gemini he could send me. I wasn't really expecting anything, but he wrote back enclosing a package of mission reports, booklets and other documents. Of course, all of it is sadly long gone. I was (and am) astonished that he would write back to a child. Amazing.

Unlike some people, I really enjoyed the Martin simulations on TV. I could not understand why TV kept cutting back to the same boring shot of Mission Control. I think there should have been tons more animations available. Of course, I really wanted a camera in the spacecraft!

For the first launch of GT-6A I was sick at home with a stomach virus, but I clearly remember the shutdown at launch (that was the word they used, not 'abort'). It was stunning and dramatic. That was the first time I saw anything really go wrong with a space flight.

Most of my own (and my friends') flying in those days was in a closet with the lights out (most kids I think used a closet) and we all assumed the cabin was dark. We had no idea about cabin lights. When glow-in-the-dark paint was invented, we painted bottle caps and glued them on a piece of cardboard for our "control panel." One day I got hold of an enormous appliance carton, and that quickly became my new Gemini (and later Apollo). I was even planning to build a capsule out of wood with two army cots, but that never happened.

I built the Gemini Revell model --twice. It taught

me a word I never knew "Khaki". I think the color was tan. I wish they just said tan.

Like Gus Grissom, I thought the Gemini was the perfect vehicle for future use, such as a quick taxi for 1 or 2 astronauts to get to and from a space station or other vehicle. I think it would have fit inside the Shuttle cargo bay... perhaps Ares 1 could have lifted it.

And the Gemini-Titan launch vehicle was America's first Art Deco rocket. Very few science people care about art, but if the rocket had been designed in 1939, it would have looked just the same."

GEMINI VII AND VIA
RICE KRISPIES BARS AND HOT CHOCOLATE

Once things had begun to settle down a bit at Complex 19 there was the thought that the fix for the problem that had caused the shutdown of the booster would be as simple as shoving that tail plug back into its socket. The rest of the job would be pad clean up and replacement of expended pyrotechnics. But, as a standard operating practice the Martin engineers immediately began to analyze the engine data from the start and shutdown and found something that would make the job far more intense. Their routine look at the data showed that the thrust in one of the engines was beginning to decay before the tail plug had fallen out!

Immediately the engineers contacted Colonel John Albert who was the man in charge of Complex 19 and the Air Force's end of the launch. He headed to the blockhouse and after looking at the data ordered the engine to be stripped down until the problem was located. The engineers worked through the night and found nothing.

The next morning, however, one engineer found the dust plug that had been left in the gas generator port. At that point it is likely that the person who had originally put that tail plug in a bit too loose and felt like the goat in this whole problem turned out to be an inadvertent hero. He had accidentally saved the mission and perhaps also saved the lives of Schirra and Stafford. Had the launch, on October 25th— when the plug was securely in place— reached the point where the bolts blew the mission may have been aborted a few feet off the pad in the form of a huge fireball.

With the engine problem solved, the launch of GT-6A was rescheduled for Wednesday December 15th. That morning my normal weekday cartoons were pre-empted by coverage of the launch preparations. Dang! How would I survive the day in Annex 3 without my morning dose of "The Funny Company?" Of course there was always the chance that Mrs. Bechtol would march us across the street to the main school to see the launch on TV. That alone would cut a good hour out of the morning. Yet, as I slogged along the sidewalk clomping in my four-buckle arctic boots with the gray winter gloom over my head the launch of GT-6 was the last thing on my mind. The temperature was just above freezing and the snow in the roads was beginning to melt a bit. I wondered if there would be enough snow for Santa to land. I was looking forward getting "The Man From U.N.C.L.E." set with the toy movie camera that turned into a toy gun.

Unfortunately for me, the launch time for GT-6A was at 8:37 Annex 3 time and the last bell rang at 8:30. There just was not time enough to get us all to hang up our jackets and line up our boots, get into our seats and take attendance and then get us all up and organized to parade over to the main school in less than seven minutes. Gemini VIA launched while Mrs. Bechtol was calling off our names and we were all answering

"Here." I never got to see it go. I did, however, get the Man From U.N.C.L.E. set from Santa that year and the Illya K. Special Secret Lighter Gun plus the Illya Kuryakin toy pistol and shoulder holster. Santa must have been totally unaware that I was flunking the third grade. Of course I never could get the shoulder holster to fit properly... perhaps there was a message there somewhere.

(MEMORIES OF OTHER FOLKS); Laszlo, San Diego, CA: "I grew up on the West Coast, so one of my main memories is just how god awful early we had to get up to watch a launch. I remember wondering why they always had to launch while it was dark in California.

After the Gemini 6/7 rendezvous, the networks announced that they were going to air the film from the spacecraft during the nightly news on a particular day. We would actually get to see what a spacecraft in orbit looked like, instead of animations like we saw during the actual rendezvous.

50 years ago, not only was there no YouTube or Facebook with the videos posted almost as soon as they arrived via datalink from space (assuming you weren't just watching the live feed), most TV sets in this country used vacuum tubes and were black & white (or monochrome in today's language). The "video" was captured in the spacecraft on 16-mm film, the spacecraft had to return to earth, be recovered by the Navy, the film removed and transported back to the US, processed and edited. So there were plenty of days between something occurring in space and the public dissemination of the video.

This time the networks worked the delay for all

it was worth, so there was a huge audience for the rendezvous film. Since we were in the monochrome majority, our entire family went over to another family's house for a rendezvous-watching party.

At the appointed time we were all in front of the TV with Rice Krispies bars and hot chocolate. The paterfamilias had adjusted the color balance (there was a tendency for the red to bloom), aimed the antenna directly toward the transmitter on top of the beach-side mountain and adjusted the tuner to within an inch of its life. All the tubes were well warmed and stabilized and Walter Cronkite came on with the nightly news.

Finally, the rendezvous film was up! Not only could you see the other Gemini floating in the blackness of space, you could see the nose of the one that was filming. You had exactly the point of view of an astronaut in space (GoPro, a la 1965). There was the Earth, sometimes below, sometimes behind the Gemini. The clouds were moving, as were the spacecraft. The horizon was curved. There was lots of junk flapping behind the Gemini. And things were black, white, red, wrapped in gold like Christmas wrappings, with blue and white behind.

That evening did more to show me more of what being an astronaut was like than any number of the books, interviews, etc. before then. Of course it wasn't the whole story, but it was the best one to that point.

Three years later, we watched a live broadcast from the Moon on Christmas Eve. Now we have social media. Can I get a refund?"

GEMINI VIII
THAT'S RIGHT WHERE YOU HAVE THEM

Gemini VI performed the rendezvous from below because orbital mechanics allow the spacecraft in the lower orbit to catch up to its target in the higher orbit simply because the lower orbit provided less distance to travel.

Although the NASA documentary made on the Gemini VI rendezvous has a soundtrack where an obviously re-recorded voice of Stafford is portrayed as being the first person to sight Gemini VII. A careful reading of the transcript shows that it was Schirra who first spotted the target.

"I see something at 12 o'clock," Schirra stated at 05:04:32 mission elapsed time.

At that moment Stafford was actually blinded by the sun reflecting off the nose of the spacecraft. He couldn't see anything including the stars, but at 05:05:05 Stafford excitedly reported,

"Hey, I think I got it! That's Seven, Wally!"

Controllers at the Ascension ground station disagreed considering that spacecraft 7 was 51 miles ahead of Gemini VI at that moment. For several minutes the crew debated as to whether or not the bright object in the sky they were seeing was Gemini VII or the star Sirius. As it turned out they were in fact seeing spacecraft 7.

Aboard spacecraft 7 Borman and Lovell soon caught sight of spacecraft 6 as it originally appeared looking like a bright star but then rapidly drew near them from below.

Just 31 minutes after the initial sighting Gemini VI was 7.7 miles from Gemini VII. All of Stafford's calculations and computer readings were coming together perfectly. Schirra was making precision burns at absolutely the correct times and using minimum fuel to accomplish rendezvous. By 05:56:01 Gemini VI was at 120 feet and station keeping with Gemini VII; rendezvous had been successfully accomplished for the first time in space.

It was an exciting moment, so much so that even Frank Borman got a little humorous when asking what attitude Schirra wanted him to maintain.

"I'm going to go ahead and put it on Inertial -Neutral here and stay right on the horizon," Borman said, "if that's what Precious wants."

By the term "precious" he was the facetiously referring to Schirra.

(See Image 12, page 163.) Flying in formation that more than 17,000 miles per hour the astronauts of both spacecraft finally got a chance to see what another spacecraft in flight looked like. One of the first things that they noticed were long, gold foil coated wires or cables protruding from the back of one another's spacecraft. Borman had initially commented about this but the crew of spacecraft 6 didn't hear the message.

"You guys are really showing a..." Schirra began, "of droop on those wires hanging there."

"… On me?" Borman replied, "Where they hanging from?"

"Well," Schirra described, "Frank, it looks like it comes out at the separation between… It might be fiberglass. It's approximately… oh… 10 to 15 feet long."

"The separation came from the booster," Borman reckoned, "right?"

"Affirmative," Schirra replied

"That's exactly where you have one too," Borman stated with satisfaction, "it really belted around there when you were firing your thrusters."

"Looks like about eight or 9 feet long and double wire," Schirra observed.

"Right," Borman agreed.

For just over six hours the two spacecraft flew in close formation as Schirra maneuvered spacecraft 6 around spacecraft 7. Borman and Lovell were conserving fuel and essentially remained in a near drifting flight attitude. Both crews, however, had plenty of film and took scores of photos and a library of movies during the exercise. At one point Tom Stafford held a sign up in his window that read, "go Navy beat Army" which was a reference to the upcoming Army-Navy football game. Stafford, Schirra and Lovell were all graduates of the United States Naval Academy while Borman was a graduate of West Point.

Schirra and Stafford flew a perfect mission and achieved the first rendezvous in space. What had started out as appearing to be a jinxed mission turned out to be just the opposite. GT-6A splashed down the following day while I was locked in Annex 3 paying no attention to Mrs. Bechtol.

Making television news history, the recovery of the GT-6A spacecraft was for the first time broadcast live on TV. From the carrier WASP by way of the Early Bird satellite and a huge antenna mounted on the carrier the events were seen live in living rooms, bars and schools across the United States.

Crewmen aboard the WASP gave the huge antenna a nickname— it was dubbed "The Big Umbrella." From GT-6A forward, every recovery of a United States spacecraft would be seen live on television with the exception of Skylab 4 which apathetic TV news executives deemed to be not important enough to interrupt game shows and soap operas. (See Image 13, page 163.)

(MEMORIES OF OTHER FOLKS) Michael; USA: "I watched the launches on our family B&W TV, if not live because of a school time launch then on the evening news. During the actual flight you needed to use your imagination since there was only audio between the ground controllers and the crew. The next phase was tuning in for the splashdown at the end of the flight. The real excitement was when the film from the flight was processed and sent to the networks, generally a day or two after splashdown. I remember the excitement of actually seeing Ed White's historic EVA, and the thrill of seeing a Gemini in flight when Gemini VI-A returned with footage of Gemini VII. Months later the NASA highlights were available on 8mm that I was able to check-out at my local library, which I would watch over and over."

GEMINI VII
SIMPLY "NO"

Of course the coming and going of GT-6A left Borman and Lovell alone once again in GT-7. By the time Schirra and Stafford had arrived in orbit with GT-7 Borman and Lovell had shattered every spaceflight endurance record. They had also sent down a library of medical data and flown the Gemini spacecraft right to its limits— yet they still had two days left to fly. The spacecraft was in drifting flight and every spare inch of space had been crammed with waste of one sort or another. They had even managed to somehow lose one of their tooth brushes and were now forced to share a tooth brush. Talk about roughing it.

For the last full 24 hours of their mission the crew of Gemini VII conducted what NASA called "positive stowage." Every little bit and piece of everything had to be accounted for and its exact location inside the cockpit had to be mapped on the ground. As always, this task was required because the exact center of gravity needed to be calculated before reentry.

Prior to the reentry there was one last medical requirement that Dr. Berry wanted the astronauts to fulfill. He and his fellow doctors were afraid that if the crew got out of the spacecraft at sea and were hoisted up to the recovery helicopter and flown to the carrier, that activity may degrade some of the medical data. The doctors wanted Borman and Lovell to remain locked in the spacecraft, where they had been for 14 days, and then have the carrier came along side and hoist the Gemini to the carrier deck where the crew could be swarmed by the doctors. After all, GT-6A had just done exactly that and Schirra and Stafford had no problem with it. To that idea, Borman had a simple reply- "NO." They wanted to be out of that cockpit as soon as the flotation collar was secured. Dr. Berry, a bit reluctantly, took Borman's answer and was happy with it. Of course there were still some doctors who were convinced that both astronauts would simply die as soon as they were back on the surface of the Earth, so what the heck.

Reentry for Borman and Lovell was normal and the crew went through all of the phases of the reentry and splashdown with zero ill effects. Bets were being lost all over the medical community when the two astronauts got off the helicopter and actually walked on their own power across the carrier deck.

Although it seems a bit odd today, when we have a permanently occupied space station with men and women in space for months at a time, in 1965 no one knew if humans could spend long periods of time in space or if doing so would have some sort of ill effects on their life spans. As of this writing, 50 years after Gemini VII, both Borman and Lovell are alive and well. Borman would fly one more time and even go to the moon. Lovell would go into space three more times including two trips to the Moon- one of which would be on Apollo 8, for eight days, with Frank

Borman... again.

I cannot help but think that somewhere along the line Lovell ticked someone off or lost a poker hand to Deke Slayton, and another long mission with Borman was pay-back. In fact, however, it was just happenstance that put the two together one last time. Originally Borman was supposed to have Mike Collins as his Command Module pilot on Apollo 8. But, a bone spur in Collins' cervical spine required surgery that bumped him off of the mission and he was replaced by Lovell who had been the back-up Command Module pilot.

GEMINI VIII
PRE-EMPT ANYTHING, BUT NOT "BATMAN!"

Command Pilot Neil Armstrong and Pilot Dave Scott had taken their Gemini VIII spacecraft and accomplished something that had never been done before— they had docked with another spacecraft in orbit. The date was March 16th, 1966 and as their spacecraft flew high over the Tananarive tracking station the crew proudly reported that they were going through the planned yaw maneuver with their Agena target vehicle secured to their spacecraft's nose.

Still a bit suspicious of what the Agena may or may not do and remembering that the previous Agena had blown up, ground control transmitted up a brief warning to the crew,

"If you run into trouble and the attitude control system in the Agena goes wild, just send command 400 to turn it off and take control with the spacecraft."

The astronauts simply responded to that bit of information and then went about their scheduled tasks. Just over six minutes later GT-8 passed out of range and the station experienced loss of signal. The next station would be the tracking ship COASTAL SENTRY QUEBEC about 15 minutes away in the Pacific.

While the crew was out of communications range,

Scott noticed the spacecraft was unexpectedly out of its proper attitude,

"Neil," he said, "we're in a bank."

Both crewmen checked their attitude balls and the instruments indicated a left roll. Armstrong used the hand controller to bring the spacecraft back into its proper attitude, but it went right back into a left bank, only now it went there much more quickly.

Soon Gemini VIII was rolling and the crew could not stop it. Remembering the ground's warning, Scott used the Agena encoder to send the command 400 and shut the systems down. He did that by way of a control device located by his right arm called an "Encoder Controller." The encoder had two moving dials, one outer and one inner stacked atop one another, with the numbers 0 through 7 printed on them. Using a rectangular knob he could align his first two desired digits, in this case 4 and 0 and then he could transmit by selecting the final digit. That was done with a rotating toggle switch that would transmit a 0 if turned left and a 1 if turned right. Scott successfully shut down the Agena's systems, but the roll rate increased and began to go off-center into a tumble.

Now, concerned that the torque from the tumbling may actually sheer the two vehicles apart and probably take the spacecraft's nose with it, Armstrong decided to disengage from the Agena. First Scott awakened the Agena once again and preset the recording device so that ground control could take control of the vehicle later— then he gave Armstrong the word that he too was ready to separate from the target vehicle.

Once the Agena's mass was gone, however, the Gemini began to tumble much more violently.

Instantly the crew realized the problem was in the Gemini and not the Agena. In short order the rates would quickly reach an astounding one revolution per second. The crew's vision began to tunnel and if they blacked out they would never awaken.

While all of this was happening on the other side of the planet, I was just settling in at our warm little house in Sheridan Park. It was just a bit after 7:00 in the evening Lexington Drive time and we had just returned from my Cub Scout "Spaghetti Dinner" that had been held in the common-purpose room at St. George's Church.

My entire after-school time that day had been taken up preparing for the dinner. Since my Dad was the troop leader, not only did I have to attend, but I had to set-up, serve and then clean up! Gee, I'd always thought that Cub Scouting was about camping and junk... but fund raising? Ugh. Still, I returned home with a belly full of pasta and a face full of tomato sauce.

Now it was time to settle in by our black and white living room TV and watch my then most favorite show in the whole world— "Batman." The show had only been running since January, but it hooked me like a drug. Everything I played somehow turned into Batman and so did everything I drew or talked about. So no one else on Earth, or off of the earth existed from 7:30 until 8:00 on Wednesday and Thursday evenings— there was just me and the TV.

The Wednesday episode was presented with its cliff-hanger, then there was the following Thursday at school when we all discussed how Batman and Robin were going to escape Thursday evening with the conclusion. There followed six days of waiting for the following Wednesday to come and another "Batman" episode. If my third grade teacher thought I was distracted before the "Batman" series came along— she was now without hope that I would ever recover.

"Batman" had just gotten started, the episode was called "The Purr-fect Crime" and Cat Woman looked quite fiendish tonight. Then, suddenly, just when things were getting good the ABC TV Network's "SPECIAL REPORT" screen popped up! What?! Not now! Not during "Batman!" What could this be? An atomic strike by the Soviets? Perhaps, but at least wait until after the cliff-hanger to tell us about it. Maybe it was just a short report and then we'll be back to the caped crusader. Nope— there was Jules Bergman breathlessly announcing that something had gone terribly wrong on Gemini VIII. Worst of all, the report eventually totally pre-empted the entire episode of "Batman." No cliff-hanger, no Cat Woman, no Commissioner Gordon, no Bat-poles, no Bat-Cave— whatever had gone wrong with Gemini VIII could have at least waited until 8:00 when that stinking "Patty Duke Show" came on.

Additionally, the news people did not know much at all about what had happened, yet they talked on and on. I was totally ticked off— I mean, pre-empt anything, but not "Batman."

GEMINI VIII
WHAT COULD GO WRONG?

Armstrong and Scott had been training together for Gemini VIII since late August of 1965. Their mission was to be a highly comprehensive flight designed to exercise the prime functions of Gemini. They were scheduled to do the second rendezvous and docking of two spacecraft and while docked Scott was to do the second Gemini EVA. His spacewalk was to be an extended two hour long ordeal where he would test out equipment such as a power drill designed to operate in zero gravity. Overall it was to be a four day mission and the crew was given seven months in which to prepare.

A lot happened in those seven months as Gemini VI saw its docking scrubbed when its Agena blew up on the way to orbit. Additionally Gemini VII and VI proved the concept and mathematics of rendezvous. Thus, the crew of Gemini VIII would now use a proven method to do a never before accomplished task with a vehicle whose previous model had blown up in space— plus they would do a spacewalk around it. Of course, they would be using the highly reliable Gemini spacecraft. What could go wrong?

Atlas 5302, the booster for the GT-8 target vehicle, was erected on Pad 14 on January 3rd, 1966 and began a routine check-out. On December 8th, 1965 McDonnell delivered Gemini Spacecraft 8 to the Cape where it was kept in storage.

Two weeks later, January 18th, Agena 5003, GATV 5003, was headed to the Eastern Test Range at the Cape for delivery, but bad weather prevented its arrival and it was not delivered until the 21st.

GT-8's booster had been erected on Pad 19 on January 13th and began testing and inspections. During that process cracks were found in the manifold of the second stage engine on the second day of February while leak checks were being done. The tiny cracks were X-rayed and found to be isolated to a single weld. Re-welding solved the problem and the systems re-work and validation was done by February 9th.

"Plan X" was conducted with GATV 5003 and Spacecraft 8 between January 26th and 28th.

It is sometimes said that there were two "X" plans in Gemini. The second was supposedly a "Plan X" which was reportedly what was written on the cover of the confidential plan for an EVA on Gemini IV. The first "Plan X" however was hatched on June 19, 1963 and consisted of a scheme to test the compatibility between a specific Agena target vehicle and a specific Gemini spacecraft. All of the codes that were to be sent from the Gemini's encoder had to be tested to ensure that each was received by the Agena and performed the function that was intended.

The only way to accomplish this testing was to take the actual flight vehicles and place them somewhere that would offer the least electronic interference. For that purpose, a "Timber Tower" consisting of a large wooden platform measuring 65X25X50 feet was constructed in an open field on the Merritt Island Launch Area Radar Range. There Radio Frequency Compatibility tests were done on every Gemini and Agena pairing before they were launched.

Yet, following the explosion of the GATV 5002, there were still suspicions concerning the Agena within NASA. This was without regard to the fact that some 180 Agenas had flown previously with only a few minor glitches. Additionally, the cause of the GATV 5002's loss was traced to a change made specifically for Gemini. As the launch day neared, NASA officials still held doubts about the Agena. Lockheed officials, who had designed and built the Agena vehicle, were far more confident. Oddly, the doubt inside of NASA would remain right through the end of Project Gemini.

GEMINI VIII
A MANEUVER WHICH NEIL
ARMSTRONG DID NOT LIKE

Launch time for Gemini VIII was 11:41:02 Nelle Haley Elementary School time on Wednesday morning. The whole event took place while I was sloshing my way home for lunch through the half-melted winter snow. Temperatures were hovering just above freezing and a gray overcast sky showed that winter was not yet willing to release its grip on the Great Lakes area.

The whole launch was over by the time I came through the front door and the news folks on our living room TV were talking about the rendezvous and illustrating it with all sorts of gadgets.

NBC channel 5 had a model train set with a Gemini and Agena on rail cars going in circles. CBS had an actual computer with an animated set of orbits all of which would have held my attention all afternoon when I was back in school except for the fact that Annex 3 did not have a TV.

Thus, I busied myself at the task of flunking the third grade. You may scoff, but it was not easy— Mrs. Bechtol was constantly on my case. One day she even had me up in front of the entire class so that she could ridicule my new Batman T-shirt and compare me to the class smart kid in order to set an example of what a failure looks like. I was often scolded for always looking out the window and she loudly chided me saying that no one would ever give me a job where I looked out the window all the time. Three decades later, while flying as an airline pilot I sat there looking out the window— and getting paid to do exactly that. I always snickered thinking about that.

Meanwhile Armstrong and Scott entered orbit successfully and began the series of OAMS burns that would allow them to meet GATV 5003. That vehicle's Atlas booster had launched exactly on time and the Agena inserted itself into the proper orbit. There was no hint of a hard start and the data being transmitted down was looking just fine. The process of catching up to the Agena would take nearly six hours.

At 05:58:58 mission elapsed time GT-8 came into range of the Hawaii tracking station. Mission Control called up with a simple question,

"How are you doing?"

"Station keeping at 150 feet." Armstrong advised.

Docking immediately with the Agena was not what the flight plan required, instead the crew was to spend some time station keeping while Scott used his encoder to exercise the Agena as the crew filmed and photographed the results. The crew was both happy and excited about the performance of the Agena in space as well as the maneuverability of their own spacecraft.

"Man!" Armstrong exclaimed, "That's great!"

" Man!" Scott chimed in, "That's really slick!"

"...it flies easy!" Armstrong spouted about the Gemini's nimble handling, "I'd love to let you do it, but… I think I better get my practice while I can."

"I'll have my chance," Scott resigned.

"Get yours later," Armstrong quipped, "Okay?"

Indeed Scott would get his chance to fly a spacecraft during station keeping "later," much later— as in three years later on Apollo 9. That was because the malfunction of Gemini VIII would scrub both his Gemini spacecraft piloting and his EVA, but he would do both on Apollo 9 in March of 1969.

Station keeping and exercising the Agena went on for nearly a half orbit and the crew observed that GATV 5003 was "as steady as a rock." While on the dark side of the Earth, both astronauts took some time to look at the Moon. Both men would later walk upon the lunar surface.

Once again in the sunlight, Gemini VIII was also passing into range of the tracking ship ROSE KNOT VICTOR (RKV) at 06:32:17 mission elapsed time.

"We're sitting about two feet out," Armstrong reported as soon as communications were established.

"Go ahead…" the RKV controller, Keith Kundel, began.

"We'll go ahead and dock." Armstrong replied.

"Roger," Kundel replied and then thought better of what he had just said, "Stand by for a couple of minutes here."

It took 18 seconds for the RKV controllers to verify that they had good telemetry from both spacecraft in order to give the final permission to actually dock.

"Okay Gemini VIII," Kundel came back on the radio with confidence, "You've got T/M solid. You're looking good on the ground. Go ahead and dock."

Once cleared Armstrong thrusted ahead and gently

eased the nose of the Gemini VIII spacecraft into the docking cone of the Agena; it was 06:33:52 mission elapsed time. Moments later Scott sent the command to make the cone rigid and the two spacecraft became a single unit. For the first time in history two spacecraft had docked.

"Flight, we are docked!" Armstrong reported, "Yes… it's a… really a smoothie."

"Roger," Kundel replied, "Hey, congratulations, this is real good."

Armstrong went on to report that the Agena was currently very stable and there were no noticeable oscillations at all. It was not only a good test pilot's observation, but also a note to soothe some of those nervous NASA managers. The calm nerves would not last very long.

Neither Armstrong nor Scott had recorded the exact moment that the two spacecraft began their un-commanded roll and they were out of ground contact as it rapidly went out of control. Estimates are that about 20 minutes of normal flight went by before all hell broke loose.

Once the crew realized that it was their own thruster and not those of the Agena that was acting upon them, they tried to quickly troubleshoot the problem.

For a moment it seemed like a stuck hand-controller. Finally, Armstrong decided that he needed to shut down the OAMS thrusters completely and work with the RCS thrusters. That simple act under the current conditions was almost super-human. With his vision blurred and tunneling and the sun flashing into the spacecraft through the windows like a high intensity strobe light and every loose object pinned to the walls, Armstrong had to reach over his head into a panel of 64 switches and flip the correct one.

Upon accomplishing that task Armstrong activated both RCS rings and immediately went to work with the hand controller. Indeed, it was not stuck and was working just fine. In short order he had the slowed the rate of tumble and was regaining control of Gemini VIII.

Once it was clear that he had control, Armstrong shut down one RCS ring in order to conserve fuel. These thrusters, located in the nose of the Gemini spacecraft were intended for reentry only and were also the only form of attitude control the spacecraft had once the adapter section was jettisoned. If they were to begin to leak or fail in some other way, the crew would be doomed to stay in orbit.

With that in mind a mission rule was written that said once the RCS was activated the mission must be terminated into the next splashdown zone. In the case of Gemini VIII, Flight Director John Hodge had to make the abort decision quickly. If they did not abort on this orbit the next available splashdown area would not come up for another 15 orbits— nearly a full day of flying.

Hodge did not hesitate— as the crew passed over Hawaii he ordered them to reenter next time around. At that same moment he screwed up my evening of watching "Batman."

NASA PAO announced the near-death experience to the media in their normal, understated manner at 7:12 PM Eastern Time,

"This is Gemini Control Houston. We've

encountered some trouble in the flight. Neil Armstrong has backed off from the Agena. He reported this trouble occurred at approximately 7 hours into the flight. We were advised of it some 5 to 6 minutes ago over the Coastal Sentry Quebec. The present situation is, he's using the rate one of his RCS rings to maintain attitude and we are watching the situation very carefully. This is Gemini Control Houston."

This woke up the hardcore space reporters who had been on the air telling the free world the victorious news of the successful docking. Of course the PAO announcement said very little about what had actually taken place and no one outside of NASA's controllers and management knew that they had come within seconds of losing the crew. All that got out was the official statement. Two minutes later PAO made another statement,

"This is Gemini Control Houston again. Apparently, the last transmission was not received by all stations, so we repeat.

Trouble has developed in the flight of Gemini 8, while docked with the Agena the configuration apparently got in a maneuver which Neil Armstrong did not like. He elected to undock and back off from the Agena.

It is also our understanding that his maneuvering system, his onboard maneuvering system, that is his big thrusters on the adapter are inoperative. He is using one of two rings in his reentry control system to maintain his attitude during flight.

It is possible that the flight may be terminated early. This trouble developed, according to Armstrong, at 7 hours into the flight. It was passed on to us some 10 minutes ago while the spacecraft was over the Coastal Sentry Quebec. We presently read 7 hours, 34 minutes into the flight and we are standing by. This is Gemini Control Houston."

In retrospect the statement that the vehicles got into, "a maneuver which Neil Armstrong did not like," is a prime example of NASA understatement. Yeah, he was tumbling at one revolution per second and near blacking out— I guess that would be a maneuver Armstrong did not like. I've not been able to find out if anyone in the 1966 hardcore spaceflight media had any indication of just how serious the situation was, but without doubt ears were perking up. The next PAO announcement came at 7:44 Sheridan Park time. It was serious enough to cause the networks to preempt "Batman."

"This is Gemini Control Houston. We are 8 hours and 3 minutes into the flight of Gemini 8.

And in view of the trouble encountered at 7 hours into the flight as reported earlier, the Flight Director has determined to terminate the flight in the 7-3 area.

We plan to bring the flight down on the 7th orbit in the 3rd, what we call the 3rd zone, which is approximately 500 miles east of Okinawa, it's in the far west Pacific.

Our situation out there is as follows: a destroyer named the MASON is about 160 miles away at this time; it is proceeding towards the point; and it should take that destroyer probably 5 to 6 hours to reach the point, which should come very close to the... well it may be a little delayed, get there after the landing itself.

The first estimate I have on the retrofire time is 10 hours and four minutes into the mission, in other words two hours from now. Landing should take place some 25 to 26 minutes later.

In addition to the MASON, a C-54 has been dispatched from Tachikawa Air Force Base in Japan, its proceeding to the point. Another C-54 is proceeding to the point from Okinawa.

The, another location here on the landing point is quoted to me as 630 nautical miles south of Yokosuka, Japan.

The weather conditions out there are partly cloudy, visibility ten miles, and the landing will be made in full daylight, its 12:30 p.m. out in the 7-3 area.

Again to recap our situation as best we understand it at this point, at 7 hours into the mission, Neil Armstrong noted some maneuvers, the spacecraft-Agena combination getting into some rates that he did not like. He elected to back off from the Agena, undock, which he did.

We do not know whether the problem developed in the Agena or in the spacecraft. We do know that the maneuver thrusters on the aft end of the Gemini 8 spacecraft became inoperative in yaw.

He did report later he had his pitch thrusters and he also reported he had used one of his reentry control system rings to bring the spacecraft under complete control.

That pretty well sums up our situation here, we'll pass along additional information as it develops. This is Gemini Control at 8 hours, 6 minutes into the flight."

That was it— this was an abort pure and simple— Gemini VIII was coming down and it was doing so in one of the emergency landing areas.

In news rooms across the free world teletypes came to life as their urgent report bells rang, telephones rang and typewriters clacked urgently— a true emergency in space was taking place and a thousand questions were left unanswered by NASA.

In the United States the three TV networks quickly took the air away from their regular programming. Not only was "Batman" preempted but so was NBC's "The Virginian" and ironically so was CBS's new hit show "Lost in Space." In fact nearly all of Wednesday night's TV line up gave way to Gemini VIII. Yet space-buffs who were older than me and far more "into it" than my scrawny Batman-crazed self would now be suspended in a mission broadcast that was garnished with doubt and questions some of which would not be answered for weeks.

Slowly NASA began to release information giving a few more details on GT-8's situation. At 8:01 Lexington Drive time PAO fed some more bits of information to the press,

"At this time the flight control team are reviewing retrofire parameters to bring the spacecraft in the 7-3 area. We have been further advised that an Grumman HU-16 amphibian type aircraft is being dispatched to the scene, again located about 506 miles east and slightly south of Okinawa.

In additional discussions with Gemini VIII over the ROSE KNOT VICTOR, it was a fairly short pass, Neil did advise that the thruster number 8 apparently failed in an open position. We don't know that that was the complete extent of the difficulty, but he did verify that that thruster failed. We don't know whether it failed while he was still docked to the Agena or after he had undocked and backed off from the Agena, but that's of no consequence at this point. He does have the spacecraft in a very stable attitude, using his RCS system.

The Agena is described as ahead and below him, which apparently is a favorable condition for the kind of landing maneuver we want to attempt, …retro command (at) 10 hours and 4 minutes into the mission and to correct an earlier statement it is approximately 32 minutes later that the spacecraft should touchdown in the West Pacific."

That was the first official confirmation that there had been a major failure in the Gemini spacecraft itself. This was a major shock to the spaceflight community because it had been thought by everyone that any trouble would most likely be in the Agena rather than the reliable Gemini.

Prior to retrofire NASA PAO made it very clear that the communications during the recovery would be sparse due to the fact everything was supposed to be relayed between the C-54 search and rescue aircraft and Japan and then sent on to Mission Control.

In fact, communications following retrofire were almost nonexistent. Immediately after the retro sequence began at 9:46 pm Sheridan Park time, Armstrong radioed back that he had fired four good retros and that was all that was heard.

Now the people in Mission Control could do nothing other than hope that their contingency recovery training and procedures would work. There was an awful lot of Pacific Ocean out there and there was no big aircraft carrier with its powerful radar to track the descending spacecraft. There was only the destroyer LEONARD F. MASON and it almost three hours away from the predicted splashdown point. Gene Kranz's team had joined Hodge's team in Mission Control for the retrofire and recovery. They had trained for this phase of GT-8 and at this point both Hodge and Kranz agreed that it was a good time for Kranz's team to take their shift— even if it meant just sitting and waiting.

GEMINI VIII
NAHA ONE THIS IS GEMINI VIII... OVER

Aboard the Gemini VIII spacecraft Armstrong and Scott went through their reentry procedures. Their actual atmospheric reentry would take place over Red China, which in 1966 was a society so closed it made the Soviet Union look like a vacation resort. If the spacecraft was to land short and come down in China, Armstrong and Scott could end up spending a few decades in prison just for dropping out of the sky.

As the reentry fireball cleared, the crew came face to face with another little anomaly in the Gemini program's planning. The reentry module was not designed to land on solid ground under its parachute with the crew aboard.

They had been trained that in the event they found themselves coming down over land they were supposed to eject in order to survive the landing impact. The only problem was that the crew's windows were pointed up while on the parachute and so there was almost no way to see if you were coming down over land or water or even over the carrier. That little oversight in the planning now became a large worry because with no tracking data the crew had no idea if water or land was below them. Some of the previous Gemini flights had ended up many miles short of their target and if GT-8 did that, it would not be a good thing in many ways.

Each Gemini spacecraft had switches located behind the astronauts over their shoulders. In order to read the switches the crew was each given a mirror on an articulated arm that was located on their main forward instrument panel.

Made by the Liberty Mirror division of the Libby-Owens-Ford Glass Company, the mirrors could be used to find these switches since the astronaut's helmets did not allow them to turn their heads that far. Additionally, the labeling for those switches was printed backward so that the crew could properly read it in the mirror.

Now Armstrong used his mirror for another purpose, he angled it up to his window so that he could get a slight peek behind the spacecraft. Much to his relief, he saw blue ocean and not a ticket to a prison camp in Red China.

No amount of training could prepare the astronauts for what they had to endure after they splashed down.

Mission Control had briefed the crew that the waves were only three feet high in the recovery area, but what was not included in that briefing was the fact that those tiny waves were on the backs of 20 foot swells! Armstrong and Scott found themselves riding an unending slow rollercoaster— and a Gemini spacecraft was a lousy boat. The reentry module rolled and bobbed insanely.

To make the situation worse, once they were on the water their ventilation system was shut down and the fumes from the scorched heat shield saturated the cabin. Armstrong soon got seasick and grabbed the spacecraft's lone sick-sack to relieve himself. Before long, Scott was also seasick and the two courageous astronauts ended up sharing the sick-sack. They did not dare to crack the hatches as the Gemini was rolling to the point where one hatch was under water and then another was submerged. Even cracking the hatch would cause a flood and the loss of the spacecraft as well as its radio locator beacon and dye marker. They just had to sit tight and tough it out.

Scott repeatedly called out over the radio to "Naha Rescue" which was the call sign for the C-54 rescue aircraft. There was no reply but he continued to call out in the blind.

After about a half hour the crew heard the sound of an aircraft overhead, but their hearts sank as the drone of the engines soon faded away. Both astronauts thought that the search aircraft had missed them and flown away.

In fact, the aircraft had dropped a team of para-rescue swimmers as well as a flotation collar. Shortly after the plane had vanished the crew saw the parachutes as the swimmers dropped into the water nearby.

The spacecraft, however, was rolling so heavily that the swimmers had a great deal of difficulty attaching the floatation collar. It seemed like an hour before the crew could open their hatches and get some fresh air. It took another two hours before the MASON came up and hoisted the crew as well as their spacecraft to safety.

By 11:17 that evening the word came from NASA PAO that the flotation collar was attached to Gemini VIII. Ten minutes later the official word came that the crew was "OK."

I was asleep and safe in my bed by then after spending the evening grumbling about "Batman" being preempted. I was not the only one who was grumbling. ABC received a flood of phone calls from other Bat-fans complaining about the loss of the episode. Likewise a cascade of phone calls came into CBS complaining about the interruption of "Lost In Space." Apparently almost no one complained about the preemption of "The Virginian" over on NBC.

(See Image 14, page 164.)

GEMINI IX
MISSED APPROACH

Landing an aircraft in foul weather is done by way of electronic aids to navigation. Today we pilots are steeped in the precision world of Global Positioning Systems and onboard navigational aids that allow a pilot to actually construct a precision approach for almost any given airport.

Instrument approaches consist of an electronic guidance that aligns the aircraft with the desired runway and another guidance that allows a safe descent clear of obstructions. If the pilot finds that the aircraft is somehow out of position or traveling too fast or if the weather is worse than reported, the pilot is obligated to abandon the approach and fly a pre-determined route that will take the aircraft to a safe altitude and position where either the approach may be tried again or the pilot can go to an alternate airport. That process is called a "missed approach" and is as critical to safety as the approach itself.

As a highly-trained professional aviator and flight instructor, I can tell you that one of the hardest things for some pilots to do is to give up a botched approach and "go missed" or execute a missed approach. This is especially true when you have the runway in sight but are simply not set up to complete the landing properly.

For those of us who have a great deal of experience in heavy weather and doing approaches down to minimums, executing a missed approach is almost second nature. You actually start every approach planning to have to "go missed," and when the approach turns out to be a good one you simply feel that much better. Once when flying for a regional airline in the Great Lakes area I did nine instrument approaches in five days and four of those were down to minimums; piece of cake.

Trouble in instrument approaches comes when the pilot refuses to let go of a botched approach. This is often the case when the pilot can see the runway and simply throws away the fact that the aircraft is way out of position. The result is what is sometimes called "chasing the runway." It is caused by the pilot not wanting to admit to themselves that they botched the approach and they then fly haphazardly in an effort to make a rotten approach good again; the results can be fatal.

On February 28, 1966 a flight of two NASA T-38 aircraft were headed to Lambert Field in St. Louis, Missouri. Aboard the lead aircraft, NASA 901, was Elliot See and Charlie Bassett— the prime crew for Gemini IX. Following in formation was NASA 907 with the Gemini IX back-up crew, Tom Stafford and Gene Cernan.

Any professional pilot will likely tell you that Lambert Field is a great airport located in the place where the crummiest weather tends to converge. Weather there can change on short notice and you need to keep yourself in tune with the trends. On this February day, the forecast was for 1,500 foot ceiling with three mile visibility— piece of cake.

There was a Notice to Airmen or NOTAM out that the glide slope on the active runway was out of service. This is an electronic beam that is transmitted from the end of the runway normally on about a three-degree upward angle. The pilot picks it up on an instrument that has a horizontal and a vertical needle. The vertical needle represents the runway's extended centerline and the horizontal represents the path that needs to be followed down (i.e., the slope) in order to safely reach the end of the runway.

By flying so as to keep both needles centered a pilot can precisely fly down through bad weather and make a safe landing.

When the horizontal glide slope is out of service the approach gets different. Now you fly with the vertical needle, or "localizer" centered and "step down." This means as you cross points, or "fixes" that are designated on your approach chart, or "plate," you are allowed to step down to a designated lower altitude and you remain there until crossing the next fix.

Normally these sorts of localizer approaches have a minimum descent altitude. Once there you fly to a final fix where either you have the runway in sight and you can make a visual landing, or you do not have the runway in sight and you are required to go "missed approach."

Along the approach, your aircraft is required to reduce to and maintain a given safe airspeed. That speed depends on the category of aircraft. The objective is to slow the aircraft to a speed where flaps and landing gear can be deployed and as you approach the final approach fix you have a speed that will allow you to safely and easily transition into a normal landing or into a missed approach. Speed control is as essential as any other factor in making an instrument approach. Often an experienced instrument pilot will have their speed exactly on the money or the approach will be abandoned and reset for a more controlled attempt. Being on-speed is essential.

In the case of NASA 901, Elliot See had three fixes that were radio stations associated with the approach- the outer marker, middle marker and the inner marker. These are small transmitters that broadcast a tone and cause a light to flash in the cockpit— usually white for outer, amber for middle and green for inner.

At the outer marker he should have had his aircraft slowed to 165 knots and started to configure for landing by dropping his landing gear and flaps as required while descending to the designated minimum descent altitude, which for his category of aircraft would have been 400 feet above the runway touchdown zone.

Since See was flying lead and Stafford was in formation, Stafford was not focused on his instruments, but had his head to the side focusing on See's wingtip.

Stafford later stated See called that he had the outer marker and both pilots put their landing gear down. Stafford later said that when he glanced at his airspeed he saw that they were flying at 225 knots, 60 knots faster than the normal approach speed. Frankly, speaking as a pilot who has flown jets in those conditions on that type of approach, the approach should have been broken off at that point. It was not properly set up and the sensible thing to do was to "go missed" and come back and set it up at the correct airspeed.

Whether it was a T-38 or a corporate jet or any other sort of jet, it makes no difference— if you start off 60 knots too fast, you have already botched the approach.

Of course, such was not the case as See continued on and descended toward the minimum descent altitude. Stafford stated that just before they broke through the clouds he saw the green inner marker light flashing but he had not seen the outer marker or middle marker lights.

The two aircraft should have been at about 300 feet in altitude and less that 500 feet from the end of the runway at the inner marker. Instead, they were more than 800 feet in altitude and as they broke out of the clouds they were more than half way down the runway! See had completely over-flown the approach.

When an instrument pilot looks down and sees that the approach is that far off it is time to swallow your pride and go missed. You never try and salvage a bad approach— it is not worth it. Elliot See decided to do the opposite and turn his botched localizer approach into a circling approach.

This is allowed under the regulations so long as you can keep the runway that you intend to land upon in sight at all times and you can keep your aircraft in a position to make a safe and normal landing.

Stafford recalled seeing See's aircraft "snaking between those clouds" before he lost sight of NASA 901. At that point Stafford called missed approach to the tower and followed the charted missed approach procedure which was a climbing turn to the left up to a safe altitude.

Meanwhile, See completed his circling but again

found himself too far down the runway to make the landing. This time, however, he found that he had committed the cardinal sin in the T-38, he got too slow in the high drag configuration with gear and full flaps. The aircraft began to sink as See went to full power and commanded the afterburners on. The right engine's afterburner lit and was at full power and the left lit and was coming up to full power when NASA 901 struck the far edge if McDonnell Building 101.

The aircraft tumbled into the parking lot and impacted the ground where it exploded. Both See and Bassett were killed in the event.

Stafford and Cernan completed their missed approach and were placed into a holding pattern by approach control. They held there until Stafford told the approach controller that NASA 907 was about to declare a fuel emergency due to his low fuel level. At that time they were cleared for another approach and successfully landed.

An investigation done by NASA and the FAA determined that pilot error was the cause of the accident. It was concluded the pilot of NASA 901 while maneuvering at low altitude in bad weather at length, "…inadvertently developed a rate of descent from which recovery was impossible."

From a pilot's point of view, however, the accident began with a botched approach and poor aeronautical decision making.

Elliot See was an experienced test pilot. He was originally trained as a Naval Aviator and later flew as a civilian test pilot for the General Electric Corporation. He had 3,200 hours in fighter/trainer type jet aircraft as a test pilot. He was also a test pilot on the T-38 before going to NASA in the second group of astronauts. See had been flying the NASA T-38s in and out of Lambert field through most of 1965 as he served as the back-up for GT-5. Aviation is unforgiving no matter how much experience a pilot has— that is why approaches are a critical tool and must be done properly every time.

GEMINI IX-A
STAGING BACKWARD

With the loss of the prime crew for Gemini IX, Stafford and Cernan were assigned the mission. Although Cernan had been training with Bassett for the pilot's side of the mission since mid 1965, Stafford had just come from his Gemini VI mission. Still, on Tuesday May 17th, 1966 Stafford and Cernan were strapped in aboard their Gemini IX spacecraft at Launch Complex 19 awaiting the launch of their Agena, GATV 5004. Atlas 5303 had been erected at Launch Complex 14 on April 4th and GATV 5004 was stacked atop it on May 2nd. The weather at the Cape was typical for mid-May; thick layers of clouds extending all along the east coast of Florida.

Countdowns for both the Atlas and the Titan II went along smoothly. At 10:12 Saginaw, Michigan time, Atlas 5303 lifted off. Although Mr. Leddy's class was watching the coverage over at the main school, Mrs. Bechtol had us locked up in Annex 3 learning how to use an encyclopedia while I, of course, looked out the window. If you looked up "Lousy Student" in the "L" volume, you would probably find my picture.

The news cameras at the Cape held Atlas 5303 in sight for just 20 seconds. Following that it went into the cloud banks and all that the local cameras could capture were flashes of the flame from the three engines. Long range cameras far south of the Cape, however, had a clear shot and the kids who were growing up with spaceflight around the country and did have a TV set in their classroom got to watch the vehicle until the 01:47 mark in the boost. From that point on the booster was obscured by clouds.

Losing the live TV images of Atlas 5303 was unfortunate, because the bird was about to put on a pretty spectacular show.

Just 10 seconds before the outboard booster engines were scheduled to cut off an electrical short circuit in engine 2 caused it to gimbal hard over to the full pitch down position. This made the entire vehicle begin to slowly tumble end over end. Since the Atlas was at an altitude of just over 200,000 feet it was above most of the atmosphere and there was very little relative wind that would normally have ripped the rocket apart in such a maneuver. As the Atlas came around about 180 degrees the booster engines shut down and the vehicle actually staged going backward! A chase plane flying above the cloud layers managed to catch the entire event on film.

After rotating a total of 216 degrees the Atlas steadied itself and began thrusting back toward Cape Canaveral before straying back out to sea. Telemetry flashed confusing data back to Earth, at one point indicating that the sustainer engine had cut off and then showing data that it was still burning but the Agena had separated. The whole mess landed in the Atlantic about 90 miles off the Florida coast. Gemini IX's launch was scrubbed.

Following the unexpected loss of GT-6's Agena back

in October, NASA learned to expect the loss of another Agena and the folks at McDonnell proposed building a sort of no-frills back-up to the GATV called the Augmented Target Docking Adapter or ATDA. This little vehicle was an ingenious and simple solution to what was thought to be a problematic Agena system. The ATDA consisted of a Gemini spacecraft's RCS section on one end for maneuvering and stability and a Target Docking Adapter and docking cone on the other end with a small instrument section in between. It was practical, simple and less expensive than a full GATV.

Like the Agena, the ATDA would be launched atop an Atlas booster and would be on hand when the next Agena failed. Oddly, the next two Agenas after GT-6 were just fine, it was the Gemini and then the Atlas that both failed.

From out of storage Gemini X's scheduled Atlas, 5304, was rolled and sent to Pad 14 for erection and preparation to boost the ATDA.

Considering that the ATDA was short, dark and not very spacecraft looking it quickly gained a nickname. The NASA people associated with Gemini affectionately called the little vehicle "the Blob." Even the people in the hardcore spaceflight news media picked up on the nickname yet it was hardly ever published.

CBS's reporter David Schoumacher brought up the subject while sitting in their Gemini mock-up with McDonnell engineer Bob Sharp, but Schoumacher called it "the Glob." He also reported that he had tried the nickname out on some of the engineers at the McDonnell plant and none of them had ever heard of it. To this day, however, Gene Cernan still calls it the Blob.

There was just one hitch in this back-up plan using the Blob. Atlas 5304 was constructed to specifically boost an Agena, which would then insert itself into orbit by way of its own PPS engine.

The ATDA had no PPS engine- other than roll, pitch, and yaw RCS thrusters it was inert. There would have to be numerous hard-wire modifications done to the Atlas booster. As the spaceflight engineers like to say, "These things are not Legos, you can't just slap a different payload on top and launch it."

Fortunately, McDonnell engineers had been prepared for this booster/ATDA mismatch and had already mapped out the exact required changes that needed to be done. Originally, the maximum time allowed for the adaptation of Atlas 5304 was 18 days. The engineers got it done in just 14 days.

While all of that was being done down at Cape Canaveral, my third grade teacher Mrs. Bechtol sat down at her typewriter after school was dismissed on the late afternoon of Friday, May 27th, 1966. She had one last teacher's duty to perform before starting her fine spring weekend. Scrolling a sheet of Nelle Haley Elementary School letter head stationary into the typewriter she rapidly pecked out a brief four paragraph, 118 word letter to my parents concerning their son that culminated with "…he will need to remain another year in the third grade."

That was it, it was official, I was indeed a failure— I had flunked the third grade. My life would now amount to nothing and at the ripe old age of nine, my future was in the toilet.

Actually, it was the best thing that could have happened to me. The teacher was right— I was not mature enough to progress and I did better in school afterward; not great, but better.

At that very moment NASA was training the first group of scientist astronauts and among them was Ed Gibson— who flunked two grades in elementary school.

I learned that failure is not the end of the world, it just means that you have to go back and try harder in order to do better. It was just a missed approach where I had to come back around and try it again. In order to cope with the big, unfair, nasty world you have to learn to deal with failure. Yet, I can still picture Mrs. Bechtol dropping that letter in the mailbox and then laughing maniacally.

June 1st was set to be the next attempt to launch GT-9, which had now been monikered GT-9A… just to keep the paperwork straight. The launch of the ATDA was scheduled for 10:00am Nelle Haley time and the Gemini was to be launched at 11:38.

This time, Mr. Leddy had convinced Mrs. Bechtol that our third grade class should venture over to the main school and watch the launch in his classroom. Of course the launch was going to take place eight minutes after the lunch time dismissal bell rang, but Mr. Leddy told us that these Gemini launches were so important that we needed to stay there and witness history. I was good with that, after all I'd be out of Annex 3 and there was TV watching involved— what's not to like?

In a very normal fashion the ATDA was launched on time and inserted into its planned orbit. In a similar manner, Mrs. Bechtol lined all of us up and had us put on our jackets. The sky was clear and blue and the temperature was a balmy 42 degrees as we marched over to the main school building.

Inside Mr. Leddy's classroom we took off our jackets and everyone found a place to sit on the floor. Up on the TV set was the now familiar image of the Gemini Titan II standing next to its umbilical tower with the big black hose-like umbilical sagging from the top of the booster.

At exactly 11:30 the lunch dismissal bell rang and Mr. Leddy told everyone that any of us who really needed to leave could go. Some of the kids who lived on the other side of busy Sheridan street had to leave because "Popsy" the crossing guard would only be on duty until 11:40. Some of my pals bugged out just because they could, but I decided to stay. After all, if these Gemini flights were important enough to preempt "Batman" there must be something cool about them.

The countdown was in a planned hold at T-3:00 when the bell rang in Mr. Leddy's room and shortly thereafter it began counting down while CBS and Walter Cronkite were in a Sherwin Williams house paint commercial. Odd how the guys painting houses on TV didn't get paint all over everything and cuss like my Dad did when he was painting our house.

By the time the launch came back on the TV it was down to T-2:04 and at T-1:34 Cronkite stopped talking and turned us over to NASA's Jack King for the final moments. King said that they were now at T-3:00 and holding once again. There was some explanation about a guidance update not loading, but that stuff was over my head- I just wanted to see the rocket go.

Finally after three tries Mr. Leddy told us all to just go home as he could see the scrub was coming.

Indeed, GT-9A's spacecraft computer was rejecting the guidance update and the mission was scrubbed for at least 48 hours. I shuffled off for home somewhat disappointed that the launch had been called off. I also regretted wasting 15 minutes of my lunch period. Unless, of course, Mom was serving that stinkin' canned chicken noodle soup. In that case I was delaying the inevitable.

GEMINI IX-A
HOLY SMOLIE!

At 8:39 am Lexington Drive time on June 3rd, 1966 Gemini IX-A lifted off from Pad 19.

Once again, I was trapped in Annex 3 as Mrs. Bechtol took the attendance and thus I missed the launch completely. Meanwhile aboard Gemini IX-A the experience was far beyond any 9-year-old's wildest imagination. Although Stafford was on his second Titan II in just six months, Cernan was quickly having his rookie feathers stripped away,

"Boy, we're really moving out now!" he exclaimed at the 40 second mark. Four seconds later he enthusiastically blurted out, "Gee! Are we moving!" followed two seconds later by, "HOLY SMOLIE!"

A dedicated Navy fighter pilot, Cernan was getting the ride of his life— so far.

At staging the crew experienced the Titan II "fireball" that they expected. More than anything that event said everything going exactly as it should. Now absent of the dead weight of the 70 foot long first stage, GT-9A initiated its guidance program and it pitched down and really accelerated.

"Needles are coming - oh, the horizon is beautiful!" Cernan exclaimed as he got his first good look at a sight that only astronauts in flight get to experience.

Just a half dozen years later he would get to admire the Earth's horizon again, but he would do so while standing on the lunar surface.

At 05:42 into the flight, the second stage's engine shut down and Cernan's stomach felt like it was going to pop out through his mouth as the sensation went from the high G loading of the second stage's burn to zero gravity in less than a heartbeat. Now floating in space, Cernan was no longer a rookie.

One of the prime objectives of the GT-9A mission was to accomplish rendezvous in three orbits, or "M=3." In that designation the digit "3" stands for the number of orbits needed to catch up to the target. Yet in a rare occurrence in NASA, the "M" does not stand for anything specific. It is just a place-saving moniker for rendezvous.

In their M=3 exercise Stafford and Cernan were simulating that their spacecraft was a Lunar Module's ascent stage coming up from an aborted landing and the ATDA was the Apollo Command and Service Module in a circular lunar orbit. Coincidentally, these same two astronauts would perform the same exercise with actual Apollo hardware while orbiting the moon on Apollo 10 three years later— almost to the day.

For the next three orbits, the crew would play a complex game of numbers as they matched attitudes and velocities with precise thruster burns. Using their onboard computer as well as numbers sent up from the ground they carefully changed their orbit in steps that would bring them up to the Blob.

After nearly four hours of intense work and concentration the flashing strobe light on the Blob came into sight.

Prior to launch the word had gotten out that Mission Control suspected the nosecone fairing on the ATDA may not have separated. On the ground, the signals were mixed and left the matter in question.

First telemetry from the rate gyros showed an expected wiggle in the ATDA at the time when the shroud should have separated. Ground controllers also got telemetry that showed that the temperature in the docking cone had started to rise which was an indication that the sun was shining on the exposed cone. Had the fairing still been in place the cone would constantly be shaded and the temperature would be much lower. Thus, many of the McDonnell engineers believed that the shroud was gone. During the launch, however, controllers did not get a signal that the shroud had jettisoned. Until the crew reached the ATDA, no one knew for sure what sort of condition the Blob was really in. The potential problem was discussed inside NASA, McDonnell, and was even broadcast on the TV so everyone knew about it.

GEMINI IX
AN ANGRY ALLIGATOR MUST
HAVE EATEN THE BLOB

A friend of mine in college lived in southeast Florida and his home, like many others in the area, had a freshwater canal that ran right through the back yard. He told me a story about the family dog, which normally resided quietly chained up in the back yard.

One morning the dog began barking continually. When he went out to see why his dog was barking he found that an eight foot long alligator had come up out of the canal and decided to sun itself in the yard.

Animal control was called and they wrangled the alligator and took it far away to be released. Several months went by and again the dog was barking incessantly. Once again the alligator had come back to take a position in the back yard. Animal control again came and wrangled the gator, this time saying that they would take it so far out in the Everglades that it would never come back after it was released.

Nearly a year passed and one morning just before dawn the dog began barking, but suddenly stopped. The next morning he went out and found the dog house, the chain and the alligator, but no dog. The angry alligator had eaten the dog.

At 03:37:22 mission elapsed time Cernan exclaimed

that he had spotted the ATDA! Stafford confirmed the sighting and Gemini IX-A continued to work their way toward the Blob. For the next 33 minutes the crew crunched numbers, double and triple checked their navigation and closed in on the ATDA. Then to their apparent disbelief they saw that the nosecone shroud was still attached!

"Would you believe that there's a nose cone on that rascal?" Stafford wondered.

"I don't know, Tom," Cernan replied, "The back of the TDA is painted, there can't be a nosecone on it, can there?"

From reading the transcript it sounds as if everyone involved in the mission other than Stafford and Cernan knew that there was the possibility that the nose shroud was still on the Blob.

(See Image 15, page 164.) GT-9A was just 780 feet from the ATDA and continuing to close the distance between the two vehicles. They were in darkness but the ATDA was illuminated by reflected moonlight that was so bright that Stafford quipped it was blinding him. As they drew to within 300 feet the crew got a clear look at the shroud problem,

"The shroud is half open. The shroud is half open on that thing," Stafford observed, "The back pin didn't fire."

"It's there but it's half open all right," Cernan agreed.

Stafford asked Cernan to turn on the docking light so they could get a better look.

"Look at that thing!" Cernan complained, "Darn it! What a mess!"

Just about that time GT-9A came into range of the Hawaii tracking station and Stafford established communications.

"I've got a weird lookin' machine here," Stafford reported.

"What's it look like?" CAPCOM inquired.

"Okay," Stafford explained, "Both the clamshells of the nose cones are still on, but they are open wide. The front release has let go, the back closing bolts attached to the ATDA have both fired and it appears that one of the bolts from the band has fired. What's keeping it together is the slip disconnect for a small electrical connector that fires the bolt on the band… So the jaws are like an alligator jaw that's open at 25 to 30 degrees. Both the piston springs look like they are fully extended."

Mission Control took more detailed information from Stafford and a scheme to break the shroud loose by Rigidizing and Unrigidizing the docking come was discussed but Stafford was too close to allow them to try that and also wanted to do it while passing over the States and being in direct contact with Houston.

At 04:32:42 mission elapsed time Tom Stafford gave Hawaii CAPCOM an iconic description of the ATDA that would live forever in spaceflight history,

"It looks like an angry alligator out here rotating around."

So it was that the nickname "Blob" was eaten by the "angry alligator." Forever more the nickname for the ATDA would be the Angry Alligator.

Although schemes for dislodging the shroud had been discussed, none seemed plausible or completely safe. It was suggested that Stafford may try hitting it with the docking bar on the spacecraft's nose. Of course no one knew how much angular impact the spacecraft's nose could handle. Plus, inside the nose was stored some fairly important things— like the parachutes.

Additionally there was a plan kicked around that would have Cernan go out on his EVA and float over to the ATDA and cut the strap with a pair of scissors that were available onboard the Gemini. This plan was being highly promoted by Buzz Aldrin and his group had tested the scissors on a similar strap and found that they would indeed cut it.

However, Cernan's EVA would be only the second one ever done by an American and only the third one ever done in the history of humanity. No one knew what that spring-steel strap would do when cut and there were no hand-holds of any sort on the ATDA. NASA management rejected that idea.

In short order, NASA management decided that actually docking with the ATDA was not required. They had proven on GT-8 that they could easily accomplish that final few feet.

Thus, GT-9A would now practice the rendezvous portion by backing far away from the ATDA and rejoining it again. Their first test was to be done without radar to see if a crew could accomplish the task with charts and hand calculations. It was a difficult brain-draining process but the crew pulled it off.

By the time that the manual rendezvous was finished Stafford and Cernan were both mentally and physically exhausted. They requested and got a sleep period of six hours.

When they awoke their next task was to simulate that they were an Apollo Command Module that had to drop down from a high orbit to rescue a lunar module stranded in a lower orbit. This was another highly taxing operation as they thrusted into a high orbit away from the ATDA and then came back down again. Cernan's two hour EVA was supposed to be next, but the crew recognized that they were still fatigued and needed more rest. Mission control agreed and allowed them to reschedule the EVA until the following day. They said bye-bye to the angry alligator and spent the rest of the day doing photography and lightweight science.

As the day came to an end the crew was tired, a bit frazzled and making little mistakes. In one of their final actions Cernan was copying a modal update and found himself scribbling.

"We should have an eraser up here," he quipped.

Yep— they could have pencils, but not erasers— because of the eraser crumbs.

(MEMORIES OF OTHER FOLKS) Ronald; USA: "I have very few memories of Gemini because I was 2-3 at the time; except for one. I have a very fuzzy memory of seeing a photo of the Angry Alligator on TV. I tried to build a model of it out of my wooden blocks with two yellow triangles. I really don't know why that is such a strong memory to me.

I also remember Gemini because I had a little Revell model that I carried around for a long time. Then I first read about the missions in Life magazine's Apollo 11 special. I haven't stopped wanting to learn about those flights since those days."

GEMINI IX
NO ONE CONSIDERED

When Gene Cernan was ready to open his hatch and egress from Gemini IX-A he was set to begin mankind's third spacewalk. Yet no one at NASA had taken the time to fully consider that fact while making the final plans for the EVA.

Leonov had done a total of 14 minutes just floating around on mankind's first EVA and nearly died trying to get back into his spacecraft. Ed White had done 22 minutes of simple floating outside of Gemini IV and was exhausted after they got their hatch closed.

Additionally, White was, without question, the most athletic of all of the astronauts at that time. Dave Scott was scheduled to do America's second EVA that was supposed to last through one night pass and two day passes but the abort of his Gemini VIII caused the EVA to be scrubbed leaving NASA with just 22 minutes of EVA experience in their logbook. Now Cernan was supposed to do a two hour long EVA that was crammed with tasks that no one had ever attempted before. No one had considered the real hazards of those tasks or the simple physics of attempting each one.

Oddly, when someone takes a basic physics class the lessons normally use a common qualifying statement to allow the basic mathematical equations to work; "assume a frictionless surface." Yet when all of the engineers who planned and simulated the Gemini EVAs did their work, they completely forgot that in space you actually have a near frictionless surface. In which case Newton's third law of motion applies to its greatest amplification; for every action there is an equal yet opposite reaction. Working in conjunction with that in space is Newton's first law of motion; an object in motion will continue in motion in a straight line until acted upon by an unbalanced force. When you do not consider the basic physics, especially in spaceflight- trouble happens.

Following White's amazing EVA everyone in NASA entered a state of EVA euphoria. "This is easy" was the common assumption simply because White made it look so easy and the EVA itself had gone so unexpectedly well. In the eyes of NASA management they had checked off one of the three main objectives of Project Gemini.

The fact that White had been exhausted with his visor fogged over after the struggle to get his hatch closed was simply ignored. That all was overshadowed by the air of "We did it!" In that atmosphere it was simple to pile tasks onto Scott's EVA and place a larger pile onto Cernan's EVA. The step-by-step learning process that had gone into most other aspects of spaceflight was simply not considered.

Cernan exited the hatch on Gemini IX-A to begin his EVA, which would be five and one half times longer than any previous spacewalk, without the benefit of a zipgun to aid in propelling him. All that he had to maneuver himself with was his umbilical.

His first task was to test "umbilical dynamics" even though Ed White had done that and stated that the umbilical was not a good way to get around during a spacewalk. Cernan was given the same task and got worse results and would later state that his wrestling match with the umbilical was like fighting an octopus.

Now Newton's first law of motion was applied to Cernan; a body in motion will tend to remain in motion in a straight line until acted upon by an outside unbalanced force.

Cernan's umbilical had been rolled up tightly inside the spacecraft and while it was stored it developed a sort of elastic memory. Once he uncoiled it in space it wanted to go back into its coiled shape. Thus every twist and wrinkle in the umbilical was constantly applying countless outside unbalanced forces that were transmitted onto Cernan. When he tried to make the umbilical go in a desired direction it came back and transmitted unwanted opposite directions to him and before his reflexes could respond, different outside unbalanced forces were being applied. The umbilical did indeed take on a life of its own and the more Cernan struggled with it the better it fought back. There was no friction to help Cernan, there were only Newton's laws and a million little lawyers applying those laws to him. He began calling the umbilical "the snake" as he wrestled with it.

"This snake is all over me." He reported to Stafford.

By the time Cernan finished his testing of the "umbilical dynamics," all that was learned was that there was no way for a free floating astronaut to control anything by using just the umbilical. In fact Cernan was completely out of control during most of the test.

Finally he managed to struggle back to the open hatch where he attempted to take a short rest. His heartbeat was over 155 and he was nearly exhausted.

It is important to keep in mind that nearly everyone thought Cernan's EVA would be just like Ed White's EVA. He was supposed to just float happily around and accomplish most of his tasks with relative ease. All of the work that Cernan had done in the zero-G training aircraft had shown that he was well able to do every task and he was in top physical condition.

No one considered that moving with the umbilical would be so draining of his stamina. Now he loitered at the hatchway knowing that the hardest job was waiting at the adapter section.

Cernan wasn't the only one having problems with this EVA, CBS News was also having a problem with EVA.

Working with the folks at McDonnell Aircraft Corporation, CBS had a full-scale Gemini spacecraft mockup constructed in their studio. Slightly in front of the mockup, a giant rotating disk had earthlike images painted upon it and spun in order to simulate the Earth below.

Inside the mockup wearing simulated G4C spacesuits were engineers Bob Sharp and Miles McClure. Overhead was the Peter Foy rig with its wires and pulleys which were attached to McClure who was supposed to represent Cernan.

The entire contraption allowed McClure to be hoisted up and appear to be floating in space just as it had done with Ed White's EVA. During both EVA's the audience heard the voices of the astronauts live while watching McClure go through all the steps planned in the EVA.

At one point, where Cernan was supposed to be working in the open hatchway, McClure accidentally leaned against his open hatch and broke it loose. The hatch door just flopped over against the side of the spacecraft mockup and hung there. Walter Cronkite immediately explained to his audience that this was not what was happening on the real Gemini. It took several minutes for the stagehands to fix the open hatch while the audience watched the moving map of the spacecraft's course.

(See Image 16, page 165.) Heading back along the adapter Cernan had additional items to test. First there was a hand-hold rail that was attached to the adapter. The second was a series of Velcro squares that were glued to the surface of the adapter's outer skin. Cernan's gloves had Velcro pads sewn onto the palms and he was supposed to use them to go hand over hand across the adapter to reach the Astronaut Maneuvering Unit (AMU). The AMU was a jet-powered backpack developed by the Air Force that was stored in the adapter section.

(See Image 17, page 165.) Cernan was scheduled to float back into the adapter, turn around and back into the AMU, strap in, power it up, exchange his umbilical connections with those of the AMU, snap the arms down and then contact Stafford by way of the AMU's radio so that Stafford could release the vehicle from the Gemini. All of that was to be done in total darkness during a night pass.

Looking back across spaceflight history it is quite easy to see that Cernan was given an impossible task.

Yet everyone in the space program in 1966 was going to have to learn that the hard way.

The Velcro on the palms of his hands that was intended to allow him to go hand over hand along the body of the spacecraft did not work. He found that the Velcro pads attached to the spacecraft simply did not stick and they came off right in his hands. He quickly gave up on the Velcro and went with the hand rail.

By way of the hand rail he got to the end of the adapter and found another surprise that no one had considered. In order to separate the spacecraft from the booster two linear shaped charges running around the circumference of an aluminum coupling ring were detonated. The ring was thereby blown into two sections and each section was left with a sharp jagged edge. Ed White had noticed this on GT-4 and even photographed it, but no one considered it.

Now Cernan had to cross the jagged edge during his EVA. Not only did Cernan have to get himself and his spacesuit over that edge, he also had to drag his umbilical across the jagged surface. A wrong move in any direction could easily have severed the line that provided electricity, communications and oxygen to his spacesuit.

Upon entering the adapter section Cernan once again came face-to-face with the laws of physics.

Although he had practiced every step in deploying the AMU in the zero-G aircraft, he had done so in very small bits and pieces each lasting about 45 seconds. Now he had to do everything in a continuous effort while floating.

He quickly found that simple tasks such as releasing and lowering the controller arms or turning valves, were nearly impossible. Each time he attempted to move anything the slightest torque of his hand or arm instantly applied the force causing his entire body to move in the opposite direction.

A series of bars had been provided to help stabilize himself. The problem there, however, was that such a stabilizing effort required that he place one hand on the bar and attempt to perform the task single-handed. Additionally, the sun had been shining on the AMU and its surroundings and had heated everything to a great degree. Cernan himself had been heated by his previous exertion and his current exposure to the sun.

No sooner had Cernan begun to work at deploying the AMU than Gemini IX-A passed into the shadow of the Earth. Two small lights had been installed to provide illumination during this night period. These were 40 watt light bulbs and only one of them was working. Cernan used a pen light to supplement this illumination. He got himself turned around and inserted into the saddle of the AMU.

Then another factor that no one had considered came into play. In the darkness of space, temperatures can drop more than 300 degrees.

Now, sweating profusely and breathing hard Cernan was building up a great deal of moisture in his spacesuit. So much in fact, that he overwhelmed the suit's ability to handle the moisture. With the extreme cold on the outside and the extreme humidity on the inside of the suit, his visor quickly formed a layer of condensation and completely fogged up. Yet Cernan pressed on and managed to switch himself over to the AMU's radio. He quickly found that the radio was nearly useless as every attempt at communication with Stafford was garbled.

Command Pilot Stafford had by this time developed a real concern for Cernan's safety. Even though Cernan was determined to keep working until he collapsed, Stafford made the "no go" decision and ordered Cernan back into the spacecraft.

The exhausted space walker made it back to the open hatch and with Stafford's help was soon inside and the cabin was pressurized.

Stafford later stated that when Cernan opened his visor his face was so red it looked like he had been in a sauna for about an hour or two too long. Stafford did the one thing that no one in a spacecraft should ever do— he grabbed a water gun and shot Cernan square in the face with cold water.

Although some critics in that era claimed Gemini IX-A consisted of a series of failures, nothing could be farther from the truth. Instead Gemini IX-A was a hard lesson in the facts of life in space. It also demonstrated just how much NASA did not know about conducting or training for EVA.

Stafford and Cernan ended their mission on a high point. They reentered with such accuracy that their splashdown was within 0.7 miles of the target. Well within view of the TV cameras and the recovery helicopters.

Their mission was a success and went a long way toward improving future flights. It just did not succeed in the ways that anyone had considered prior to the flight.

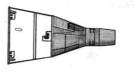

GEMINI X
MUSICAL FUEL TANKS

On September 21st, 1965 the second stage fuel tank for Gemini Launch Vehicle (GLV) 10 arrived by rail at the Martin Baltimore plant. The production rate of the Titan II launch vehicles was at its peak. Vehicle components were being received and checked out at a pace that, with any luck, might just keep up with the launch rate of Gemini missions.

Upon inspection, however, it was found that a lead acid battery had leaked during the trip east and corroded the dome on the stage's fuel tank. Martin engineers rejected the tank and ordered it shipped back to the company's plant in Denver. That posed a huge problem in the assembly and shipping schedule. It could be months before the tank was returned.

The solution to the problem led to a game of musical second stage fuel tanks. As luck would have it, the second stage fuel tank for GLV-11 was just four days from the completion of its final assembly in Denver and could be at Baltimore by November 3rd.

The decision was made to swap the tanks and fly the second stage fuel tank for GLV-11 on GLV-10. Of course that left GLV-11 without an assigned second stage fuel tank. So, the second stage fuel tank for GLV-12 was scheduled to fly on GLV-11 and in turn

GLV-12 would get GLV-10's second stage fuel tank. It was a paperwork nightmare, but it protected the schedule.

GLV-10's first stage was flown on schedule to Cape Canaveral on May 18th 1966 and the second stage containing the fuel tank for GLV-11 was flown in two days later. Both stages were transported to the Cape's Hangar L and were purged with dry nitrogen and stored under controlled access.

The scene was now set for GLV-10 to boost astronauts into orbit around the Earth.

John Young, who had flown as pilot on GT-3, would fly as command pilot on GT-10. Mike Collins, a rookie, would fly as pilot and would perform one of the most challenging EVAs to ever be attempted in the Gemini series.

In the shadow of all the issues that cropped up during the flight of GT-9A, it was important that everything worked according to plan on this next fight. Yet none of those issues altered or lessened the workload and challenge of Collins' EVA one bit.

(See Image 18, page 166.) Launch day was July 18, 1966 and it was 81° at Cape Canaveral as GT-10's Titan II waited on pad 19. This was scheduled to be an evening launch because one of the prime objectives of the GT-10 mission was to rendezvous with the Agena that had been left in orbit by the aborted GT-8 mission.

In order to match the orbit of that vehicle, GT-10 had to be launched at 5:20 PM Cape Canaveral time. For space-buffs who are growing up with spaceflight this evening departure meant two things. First it pushed most of the primary events back into the late hours of the night. Secondly, it also meant that the launch would take place during dinner time in the Eastern time zone and many families in 1966 felt that dinner time was family time and it was no time to watch TV.

(MEMORIES OF OTHER FOLKS) Mike, Dayton, OH: "I remember the TV coverage of Gemini X, I was 14. It was summer time and I had just come in from playing baseball. Hot and sweaty, I plopped down in front of the family TV and watched the news people talk about what was going to happen.

They used a toy train to demonstrate how the Gemini was going to meet up with the Agena. I thought that was really cool because trains were one of my big things.

Then after about an hour of coverage that included the Atlas launch they were getting ready to launch the Gemini. That was just when Mom rang the dinner bell and told me to turn off the TV. I tried to protest saying that they were going to launch the astronauts, but Mom wouldn't hear any part of it ordering me to shut off that darned TV and go scrub my face and hands for dinner.

By the time I finished dinner and did my after dinner chores the launch was all over. It was the only Gemini I ever missed."

Countdown and launch for Gemini X was smooth. This was Young's second ride on the Titan II and Collins' first. Throughout the first stage burn everything was perfectly normal. Although the skies over Cape Canaveral appeared clear blue there was actually a thin cloud layer at high-altitude.

"And here come the clouds," Collins quipped as he

glanced at his window.

An instant later GT-10 punched through the clouds and headed for space.

"Nice ride," Collins exclaimed.

GT-10 began to vibrate and aerodynamically roar. It was a sound that could be both heard and felt in the cockpit. Suddenly as the launch vehicle passed through the speed of sound the ride began to smooth out.

"There," Collins said over the intercom, "she's getting a little smoother. Must be supersonic there, John."

"Yes," Young replied, "that's it."

At 02:35 into the boost the launch vehicle staged. Most Gemini watchers had gotten used to the orange flash and sparkling bits of metal resulting from the fire in the hole staging of the Titan II.

GT-10, however, was different because this time the staging event ruptured the first stage's oxidizer tank resulting in an explosion that destroyed much of the stage. This meant little to the crew who were already accelerating away from the event. Yet some spectators watching on TV at home, including Collins' wife Pat, thought the entire vehicle had blown up.

"Got a bright flash over the nose," Collins reported back to mission control, "just instantaneously."

As the second stage began to initiate guidance the nose of the vehicle tilted over and the crew got their first good look at the Earth.

"I'll be," Young exclaimed, "here comes the world! Isn't that beautiful!"

"Gosh darn, that really is!" Collins agreed in a family-friendly way, "I promised no swearing, didn't

I?"

At 04:23 into the boost Young again took some time to look out through his window,

"Oh, my!" Young blurted out, "look at that... Isn't that something?"

"It really is," Collins agreed.

"That makes it worth it," Young continued, "doesn't it babe."

"Boy, that is incredible!" Collins concurred, "Look at that!"

"Put your head back in that cockpit." Young chided as both astronauts laughed aloud.

At 05:36 into the boost, GT-10's second stage engine shut down. Just 17 seconds later the crew felt two distinct "bumps" coming from the stage as the engine "burped."

Gemini X was in orbit.

GEMINI X
A MISSION OF FEW WORDS

In the Gemini era, NASA had the tendency to insanely cram more work into a mission than the crew could possibly handle. That was very much true with the mission of Gemini X as the crew had been given four days worth of work to do over a period of two days 22 hours 46 minutes and 39 seconds.

Young and Collins were scheduled to rendezvous with an Agena on their fourth orbit, use its engine to boost them to a record altitude, perform a stand up EVA, rendezvous with the second Agena, do a second EVA to that vehicle and in between all of that run a laundry list of 11 complex scientific experiments. The crew knew before they departed the pad that they had been loaded down with at least four days worth of work crammed into the mission.

As a direct result of that, the crew of Gemini X immediately turned into the least talkative crew ever to fly in space. Unfortunately, their mission was front page news not only in the United States but around the free world. Reporters sat in Houston with their pencils sharpened, their pads of paper and typewriters at the ready and deadlines looming overhead while almost nothing was heard from the crew.

In fact, while going over transcripts of mission

communications you find page after page of crew transmissions consisting of little more than "Roger" or "10 go" or other similar replies consisting of six words or less. By the time the mission had entered its second day the people in the spaceflight news media were ready to blow a gasket.

Although the simple fact was that Young and Collins were loaded down with tasks such as attempting a rendezvous guided by sextant, charts and hand calculations, such things do not translate into good newspaper copy or headlines. You cannot sell papers with a blazing headline that reads, "GEMINI CREW SAYS ROGER 287 TIMES."

Likewise, television audiences want to know what is happening as told by the crew rather than NASA PAO. Yet what they were seeing on their TV sets were simulations with space-suited engineers bumping around mockups while static crackled in the background garnished with a lot of "Roger."

TV news anchors, such as Walter Cronkite, did their best to explain that John Young was one of the least talkative astronauts in the program and at the moment Mike Collins was the busiest astronaut in the program. Such explanations didn't go far with network news producers and soon NASA began to feel the heat.

At the beginning of their second day on orbit as Collins was preparing for his standup EVA as Gemini X came into range of Houston and a familiar voice came over the radio. The crew recognized it as their boss, Deke Slayton,

"… You guys are doing a commendable job of maintaining radio silence. Since the French stopped shooting at you why don't you do a little more talking from here on?"

This subtle yet direct order from the boss said far more than the 29 words of its composition.

Number one, it helped to satisfy the press. Number two, it helped to remind the crew that the entire world was listening. Number three, it ticked off John Young.

The crew had spent the entire previous day working a complex rendezvous method that had been dreamed up by some eggheads on the ground. This pointless exercise called "Module VI" had caused aggravation and cost fuel.

Now the crew was in the process of running a 131-step checklist preparing for Collins' EVA. Any distraction could cause a missed item on that checklist that may also cost the astronaut his life. But NASA management is concerned that the crew isn't talking enough?

The crew felt they had been sent there to accomplish a complex mission in space and that did not include narration.

Yet the dynamic existed that an astronaut's job included both. It was a very difficult lesson for members of the astronaut corps to learn. Not until Apollo 10 would that larger lesson take root— interestingly enough, John Young would be on the crew of that mission.

Now, during the mission of Gemini X, the concept seemed excessively pointless to both Young and Collins. Although Collins picked up his conversation and sent down a little more narration, Young remained an astronaut of very few words.

It would not be until his flight of Apollo 16 and his

extended walks on the moon that Young would find this gift of gab.

Yet he remained a person able to get the most mileage out of very few words. No space-buff will ever forget John Young handling the press prior to the flight of STS-1, the first orbital space shuttle flight. When asked a long-winded question about the crew's ability to eject during the burn of the Solid Rocket Boosters, Young simply replied,

"You just pull the little handle."

GEMINI X
MAJOR ACCOMPLISHMENTS AND UFO BAIT

Although the crew of Gemini X never did get as talkative as the media would've liked, when their spacecraft splashed down in the Atlantic Young and Collins had achieved a number of milestones in the effort to get to the moon.

They had successfully docked and remained coupled to another space vehicle. They also used the power of another space vehicle to propel them to a world record altitude of 474 miles. Additionally they had performed rendezvous with an inert spacecraft without the benefit of rendezvous radar, done an EVA to that spacecraft and returned material from it.

Also, Gemini X was the first time that two EVA's had been performed on one Gemini mission. Even though some of their assigned tasks, such as Module VI and color patch photography had not been completed, the majority of their experiment burden had been completed successfully.

Along with all of the real science and true engineering that had been accomplished during Gemini X, one additional unintended and unwelcome product had also been produced. The mission would provide material for UFO crazies in the years ahead.

Too often, simple observations made by astronauts

are picked up and overinflated by folks in the UFO community. Additionally common terms used by fighter pilots are taken from transcripts and blown out of proportion in order to prove that what has not been seen has actually been seen or that some sort of a conspiracy exists. Unwittingly Young and Collins aided this insanity during Gemini X.

The incident took place toward the end of their first orbit while in touch with the Carnarvon, Australia station at 00:59:32 ground elapsed time. Something caught Collins' eye,

"Hey, I see stars! No, I see a bogey. Look out here John."

It should be kept in mind that fighter pilots commonly call anything that they have yet to identify a "bogey." This is not some sort of secret code language used by astronauts to describe an extraterrestrial spaceship.

"You get your head back in that cockpit!" Young replied.

"Look!" Collins insisted, "there is a light up there. John, can you see... Can you yaw around, there is a light."

"We don't have any trouble... We don't have any time to fool with it now." Young commanded.

"I see a big bright light out there," Collins went on, "I tell you John! Mark the time. Is the voice tape running?"

"Yes." Young answered.

"Okay whatever time it is here." Collins began.

"You don't have to even look at it." Young instructed. (The "it" that Young is referring to is not what Collins is looking at out the window, but rather it is the time hack on the recording. The onboard tape carried its own time hack so there was no real need for Collins to state what time he was making the recording.)

"Where I am," Collins continued noting the time as best he could, "we got a bright object. Looks like a planet or something, looks like it's in orbit with us. Looks like it's in a higher orbit and going the same way."

"It's the booster." Young stated flatly

"Now it's just… Now it's still ahead of us; it's ahead of us slightly."

"Well," Young pondered apparently now looking at the light, "if the booster is lower than we, it's off…"

"… Cause you know John is a meany and won't let me look, but it's out at high, one o'clock high, I have us in higher than we are," Collins detailed.

"Okay," Young resigned.

"And all I'm doing is pushing 95 on voice track," Collins added.

"Okay," Young said.

"It went up to about 145. Now it's back down. Oh golly, that's incredible!" Collins was fascinated. "I can't get over the beauty of it."

"Don't you look out that window!" Young attempted to get Collins attention back into the spacecraft and his work.

(Moan) "what a slave driver, John… Sun's in my eyes. I need the window shade, that's what I need, believe it or not. … I say. The sun's in the eyes and you can't push out the…"

"I love the…" Young said, "come back down to a 100. See it went up… Started at… went up to 145."

"It did?" Collins pondered.

"Yes, so beautiful." Young said, "… It looks like until we get…"

"Yes." Collins agreed, "baloney to me."

Although the crew got back to their jam packed schedule of rendezvous preparation the object came back 40 minutes later as Collins' work required him to look out the window and pick a star,

"First star…" Collins said to himself, "I have two beautiful stars right out my window. You have two?"

"No," Young replied, "I don't have any. Those are…"

"What the heck are they?" Collins asked his commander, "Did you see that?"

"I bet those are the Agenas," Young determined.

"What?" Collins asked.

"Yes." Young reasoned, "Those have to be the Agenas. We're in the same orbital path with them."

"Well, man, look at those!" Collins pondered, "They're beautiful! What's that flashing on the right? Is that the Agena? See it flash? What? John, do you see what I see?"

"Yes," Young agreed.

"What the heck is that?" Collins continued to wonder.

"I don't know," Young stated flatly.

"I see," Collins said, "is a tape recorder still running?"

"Yes," Young agreed, "I don't know what those are."

"Okay," Collins dictated into the tape recorder, "They're at 12 o'clock; but time on it is 01:39:00 and they are directly ahead. They appear to be coaltitude there are two extremely bright stars – I'd say about 1 degree apart. The one on top is slightly to the left of the one on the bottom, and about a degree below those two there is a flashing light, a similar magnitude light which is visible all the time and which periodically flashes. There went the flash, Mark; flash, Mark!"

"Oh I can't…" Young now pondering, "there's no telling what they are."

"It's really weird," Collins went on, "and when I first saw those I could see sunlight; I could see the earth lit by the sunshine down the lower right of the windows in the spacecraft nose is still in the light. And I see these two bright objects get ahead. See those, John."

"Yes," Young concurred, "think we ought to report to the ground?" Then he had second thoughts, "no, let's not." Three seconds pass… "Those aren't stars. Can't be."

"When you report it?" Collins suggested, "Tell them what you see. See what they say."

"This is Gemini X, Houston. We have two bright objects appear in our orbital path. I don't think they're stars; they look like we are going right along with them."

Houston replied with a simple "Roger" and four minutes later re-contacted the spacecraft, but only to tell them that their Module VI rendezvous solution was a "no go." Apparently, however, Young's report kicked up a little bit of dust. It took 28 minutes before mission control responded to the report of the strange lights.

"Where the objects from you?" Houston asked.

The transmission to the crew was garbled on the receiving end and while Young and Collins tried to figure out what the message was, Houston continued,

"If you can give us a bearing, maybe we can track them down."

Four minutes later Collins began to talk to Young about his sextant,

"John, you want…"

"They just disappeared," Young radioed to Houston "I guess they were… They were satellites of some kind."

To any UFO enthusiast out there those communications are like striking gold. Indeed the crew of Gemini X had seen an object that was flying and they could not identify it. This would fit the definition of a "UFO" in its most basic term.

Of course there are those who would take this basic befuddlement and blow it hugely out of proportion. The UFO conspiracy wackos enjoy taking the facts that the crew made this report and were highly un-talkative during the mission plus they returned with almost no photos of their activity and surmise that the "real mission" of Gemini X was to fly up and have a secret meeting with aliens that NASA had pre-arranged. Nonsense.

Normally persons making such claims also just happen to have a book or a video to sell or a great desire to see themselves on television.

Simple logic would ask the question; if NASA was indeed hiding such a secret mission why did they not scrub these conversations from the public transcripts of the mission?

Unfortunately, Young and Collins, like nearly all of the astronauts who have flown missions in space, would have to spend the rest of their days wondering when the next UFO wacko is going to approach them and begin asking what the aliens are like. During the flight of Gemini X Young and Collins were being very careful not to swear yet unwittingly planted bait for the UFO wackos.

GEMINI X
WRAPPING UP A SILENT MISSION

Two EVA's were contained in the Gemini X flight plan, the first being a standup EVA and the second being a free-floating EVA and a visit to the Gemini VIII's abandoned Agena target vehicle. Both EVAs were scheduled to take place in the early evening Sheridan Park time— right when folks were watching their TVs.

Unfortunately for most space-buffs the standup EVA was scheduled to take place on the dark side of the Earth. That equated to the event playing out in an area where there were only remote ground stations and thus little direct communication with mission control.

The free-floating EVA was scheduled to take place in the same area of sparse communications strictly due to orbital mechanics. Another problem was the fact that the crew of Gemini X would not be using the VOX communication system that had given Gemini IV and ground controllers so much heartburn. So, whenever Collins was talking the only way that anyone on earth could hear what he was saying was if Young pressed his own push-to-talk button inside the cockpit. It was a simple activity that the command pilot simply didn't bother with.

All went well during the standup EVA as Collins found his body positioning in his assigned tasks to be well within the area of his training.

Throughout the night pass, Collins dutifully performed each and every one of his tasks. But, as the spacecraft came into sunlight, suddenly Collins felt his eyes begin to water. In fact, they watered to the point where he could almost no longer see.

Feeling that the sunlight itself had something to do with the eye irritation Collins pulled his neck down into his spacesuit like a turtle shielding his eyes from the sun, but that didn't help. At the same time he reported this problem to his command pilot and after a brief discussion discovered that Young was also suffering from eye irritation to a great extent. The problem got so bad that the command pilot decided to terminate the EVA.

Ground controllers in Houston worked through the night attempting to pinpoint the cause of the eye irritation.

Although some thought it may be related to lithium hydroxide finding its way into the oxygen flow, the fact remains that the exact cause was never determined.

The following day the crew used their spacecraft's thrusters, what little fuel they had remaining, and sextant calculations to rendezvous with the Gemini VIII Agena. That vehicle was now out of power and completely inert. The task for Collins would be to float over to it and retrieve a small solar wind experiment.

Collins exited the spacecraft, struggled briefly with a nitrogen line to power his zip gun and then took his first try at reaching the nearby Agena. Young had maneuvered the spacecraft as close as he dared to the inert Agena and Collins pushed off from his open hatchway, floated to the Agena and grabbed the lip of the docking collar. Now, however, he discovered the same problem that Cernan had experienced. Simply grabbing the collar with his hands did not negate the inertia of his torso and legs and he quickly torqued around until his hands were simply pulled away. He floated away on his tether arcing around above and toward the back of the Gemini spacecraft.

Collins decided it was a good point in the process to use his gun to attempt to stabilize himself and set up for another try at the Agena. Now the spacewalker flew himself back to his open hatchway and decided to make another try at the Agena— this time using his zip gun rather than pushing off and leaping. Steadying himself and carefully aiming the gun Collins now flew slowly back to the Agena.

This time he reached behind the docking collar and grabbed some wires to act as a makeshift handhold. Slowly he worked his way hand over hand to the point where the solar wind experiment was located. Pushing the two release buttons and removing the small shroud he recovered the experiment.

Upon returning to the cockpit, he handed Young the recovered experiment and then began preparing to do a series of maneuvers at the end of the tether in order to demonstrate the use of the gun.

By this time, however, mission control decided that the crew had used their limit of fuel for this phase of the flight. They were in fact down to 874 pounds of OAMS fuel.

The EVA was terminated and no one on the ground heard any of it. In fact, the first transmission that ground stations heard from Collins and relayed to the media and the public took place as Collins was standing in his seat again preparing to do the gun evaluation.

"Okay, Houston, this is Gemini 10. Everything outside is about like what we predicted only it takes more time; body positioning seems to be a problem. Although the nitrogen line got connected without too much of a problem.

"When I translated over to the Agena I found that the lack of handholds is a big impediment. I could hang on but I couldn't get around to the other side which was where I wanted to get around to.

"Finally I did get around to the other side and I did get the S-10 package and the nose bearing and John now has them, however, there's a piece of shroud hanging… or a part of the nose of the Agena that came loose and I was afraid I'd get snarled up in that, and so was John, so he told me to come back.

"So, the new S-10, which I was going to put on the Agena, I didn't and I'm just now throwing it away. I also lost my EVA Hasselblad inadvertently, I'm sorry to say. I'm getting ready now to do some gun evaluation." And that was that for the sum total of EVA information sent down by the crew.

After Collins described his trip to the Agena, Houston canceled the remainder of the EVA.

To say that the news media was not happy with the information that had not been passed down from the crew during their near silent mission would probably be an understatement.

To make matters worse, John Young had problems with his camera not working properly and Mike Collins had actually lost his camera while doing the free-floating EVA, so the crew returned almost no pictures or film of their activity other than the scientific ultra-violet images and color chart experiment images.

Oddly, Collins had designed a special bracket that would make it easy for him to remove his EVA camera where using only one hand would allow him to quickly take amazing pictures during his EVA. He took a lot of photographs using that camera and bracket. It was that bracket, however, that allowed the camera to work its way loose and pull free from its lanyard then simply float off on its own orbit.

It likely reentered several months later which was somewhat hard on the film inside.

In the end, the crew made up in a small way for their lack of newsworthy information output by reentering so precisely that they splashed down well within view of the recovery carrier and TV cameras.

GEMINI XI
A PERFECT MATCH FLYING TOGETHER IN SPACE

Professional pilots always feel better flying with someone that we know very well. In the airlines and the military there are thousands of other pilots but when you were assigned to a given company, a given agency, or given squadron it's always better to bring aboard pilots that you have flown with before.

Personally, at every airline and company where I flew, I always attempted to bring in other pilots with whom I had trained, or flown with. Not only do you want to give them the opportunity to work where you work, you also know that when things get tough the person working with you can absolutely handle every situation with the degree of competence that's going to make your job safer and easier.

When NASA was selecting the Mercury astronauts, Alan Shepard and Wally Schirra had encouraged Pete Conrad to put up with the crap required to go through the selection process.

In the end Conrad got his fill of the week-long medical torture that the candidates were being put through for Mercury. He elected to go back to his flight test squadron and his sweetheart F-4 Phantom jets.

When the second round of selections for astronauts

was about to begin, both Shepard and Schirra sought out Conrad in order to urge him to try again. They had flown with Conrad and they knew him well – they also knew that he was a perfect fit for the astronaut corps.

At first, Conrad wanted nothing to do with another week of medical torture, but his two astronaut pals assured him that the process had been changed. Conrad, of course, made the cut and became one of the "new nine" astronaut group.

When the next round of selection came along Conrad himself had a candidate in mind that he knew would be perfect to fly in space with him. That candidate was his friend and fellow F-4 Phantom jock Dick Gordon.

Not only were Conrad and Gordon a perfect match to work together, they were also the perfect match flying in space together, so much so that they would fly together on two missions, one of which would take them all the way to the moon.

On September 20th, 1965 NASA announced the crew for the Gemini VIII mission. Commander Neil Armstrong, pilot Dave Scott, backup commander Pete Conrad and backup pilot Dick Gordon were selected for the mission.

Being selected for a backup crew meant that Conrad and Gordon were now in Deke Slayton's mysterious crew rotation. That rotation normally meant they would serve as backup then skip two missions and become prime crew. Thus, doing the math, Conrad and Gordon were now in line as the prime crew for Gemini XI.

The moment Gemini VIII splashed down Conrad

and Gordon went into intensive training for Gemini XI. And this is where an interesting dilemma in the human factors of Project Gemini would take place that would have an ill effect on Gemini XI.

Although people outside of the program would think that a returning crew would sit down and fully brief the next crew on what to watch out for, that was not taking place. What was taking place within the program was that the pace of missions was so intense that the individual crews were not learning the lessons of previous crews.

A good example of this involved Gene Cernan's problems during his aborted EVA. Cernan had effectively exhausted himself and nearly worked himself to death due to a lack of handholds and foot restraints and also overheated his spacesuit to the point where it could no longer handle the moisture being generated by his exertion.

Of course he returned to the ground and was duly debriefed by NASA management. But the hard lessons from his EVA were not integrated into the training of future space walkers.

During the period between Cernan's EVA on Gemini IX and Gordon's Gemini XI mission, Gordon was intensively training for his EVA. However, so intense was Gordon's training that no one took the time to brief him on the details of Cernan's problems. Years later, Gordon would recall that he knew nothing of Cernan's problems with the tether until he read Cernan's 1999 autobiography, "Last Man on the Moon."

In his 1999 NASA Oral History, Gordon put it this way,

"To be perfectly frank about it… We didn't have enough time really to assimilate the information, or we didn't grasp the significance of the information that was being brought back.

Now, remember: we were on 2-month launch centers. It was almost like just passing each other to and from the pad. And there wasn't a great deal of opportunity to assimilate the information from previous flights.

So, I don't think we did a very good job of transmitting to each other the things that were happening. And in the training for XI, I certainly didn't anticipate any problems with the EVA. As a matter of fact, the training we did was basically in a zero-g airplane."

So it was that on September 12th 1966 two good friends, Conrad and Gordon were strapped in to their Gemini XI spacecraft and prepared for launch. The two Navy test pilots waited for launch as if waiting for a cat shot from a carrier. They felt sure they were completely prepared for the mission ahead; they were mostly right.

(MEMORIES OF OTHER FOLKS) Mike; USA: "I must have been around 7 or 8 years old, I remember watching at least two of the Gemini launches on our black and white TV. During one, my dad came over and said "look, see how the tower falls away at the last second" or some such thing. He named the tower but that went right over my head. Still, I was captivated by the excitement of the launches, and wanted to know more."

GEMINI XI
WOULD YOU BELIEVE M=1?

The primary goal for Project Gemini was to prepare the way for Project Apollo. Of course, part of the primary objective of Project Apollo was not only landing a man on the moon but it was also returning him safely to the earth.

That "returning him safely to the earth" part required rendezvous in lunar orbit. In case of an emergency that rendezvous had to be accomplished on the very first orbit, or M=1. The primary goal of the Gemini XI mission was to demonstrate that this sort of rendezvous could be accomplished. In order to do that the Titan II launch vehicle had to liftoff at a precise time.

While other rendezvous flights had between 30 and 35 seconds for their launch windows, Gemini XI had a two second launch window.

Prior attempts to launch Gemini XI had taken place on September 9th and September 10th.

On September 9th a pinhole leak in the first stage oxidizer tank of the Gemini XI booster was discovered and needed to be repaired.

On September 10th problem was not with the Gemini XI booster but rather with the Atlas booster for their Agena target vehicle. On that morning the

Atlas' guidance system had a nervous breakdown and needed therapy and counseling. Eventually a brain transplant fixed the problem and the launch date was set for September 12th.

Although the launch was scheduled to take place at 9:42 Nelle Haley elementary school time, I would once again be deprived of watching it on TV. Having successfully flunked the third grade, I had spent the summer zooming around on my bicycle and pretending to be Batman complete with a bath towel cape tied around my neck.

Sometime in the final weeks of my summer vacation the postman delivered news to my home that would completely ruin the next 10 months for me.

Not only had I been reassigned to again suffer the third grade in Annex 3, but I would be lorded over by the infamous Mrs. Shea. She was a woman who probably had been passed over for the stage production of the Wizard of Oz and the part of the Wicked Witch of the West simply by the fact she so resembled the actual character that creeped out the producers. She carried a business card that said, "Have Yardstick — Looking For Knuckles" and now she had me in her sights. For some reason she seemed to gain glee from boisterously telling me, "What's the matter? This is all review for you."

I never did figure out how she was able to come up so quietly behind me in those orthopedic shoes and gain my attention with the smack of her yardstick.

Yet there was one big advantage to being in her class; she loved to nap. Nearly every day in the afternoon she would crank up the projector loaded with a movie, and we're talkin' the 18-inch-diameter-reel sized movies, and let them play while she put her head on her desk and snoozed.

Unfortunately, Mrs. Shea never liked to walk over to the main school. Thus, when there was a Gemini launch, I was again stuck in Annex 3. Such was the case on Monday morning September 12th as the Gemini XI booster roared to life and left pad 19 just one half of a second into its launch window. Conrad and Gordon performed all of their burns with precision and caught their Agena target vehicle over Hawaii. Over California Conrad radioed down,

"Would you believe M=1?"

Conrad then eased the nose of spacecraft 11 into the docking cone of GATV 5006 one hour and 34 minutes into the mission.

The fundamental steps to achieving a lunar orbit rendezvous mission had now been accomplished. Thereafter both Conrad and Gordon took turns practicing docking their spacecraft with the Agena. Gordon achieved a Gemini project first by becoming the first astronaut to accomplish a docking from the right seat.

Of course, the Gemini mission was on the TV when I got home for lunch. Mockup simulators and the pretend astronauts were pretty good to watch while Mom served up one of her lunchtime specials. More than likely it was her fried macaroni which was a standard item on the "four days until payday" menu.

Returning to Annex 3 I saw that, much to my delight and relief, Mrs. Shea had not one but two big movie reels sitting on her desk. Oh yeah, were talking a major nap for the old battleaxe.

History will show that as the crew of Gemini XI

ignited their Agena in order to make their out-of-plane burn, on the other side of the planet in Annex 3 Mrs. Shea was soundly snoozing while I busily drew yet another cartoon about her on a brown hand towel smuggled from the bathroom.

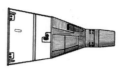

GEMINI XI
MRS. SHEA WOULD HAVE BEEN PROUD

Normally I was awake each morning well before dawn and on the morning of September 19th the crew of Gemini XI had beaten me to an early morning rise and shine. Conrad and Gordon were up and active just after two o'clock in the morning Sheridan Park time.

Today was the day that Gordon had been training long and hard for so many months. This was the morning of his first EVA and one of the critical operations of that EVA would be to move out to the docking adapter on the Agena and fasten a nylon lanyard to that vehicle.

Gordon had practiced this move in zero-G aircraft to the point where he was able to get there, straddle the nose of the Gemini, stick both feet in between the Agena's docking cone and the adapter, thus freeing his hands and allowing him to fasten the lanyard within about two minutes. Almost completely unaware of the details of Cernan's difficulties Gordon was sure that this was going to be a quick and easy.

By the time I woke up, clicked on the black and white living room TV, and dumped out by hand full of plastic spacemen onto the living room carpet, Conrad and Gordon were nearly finished with their EVA checklist.

Gordon was sealed inside his inflated spacesuit and the crew found that they were nearly an hour ahead of schedule. Conrad told mission control that considering how advanced they were on the checklist he felt it would be wise to just sit tight and rest until the clock caught up to them. Mission control agreed that it sounded like a very good idea. It would turn out to be a huge mistake.

In the wrestling match to get equipment on Gordon had already become overheated. Now a very simple problem was about to turn into a monumental struggle.

Gordon had donned all of this EVA gear with one exception— he had yet to attach the gold visor onto his helmet. In Earth training conditions that visor just snapped in to two pivots one on each side of the helmet.

The problem was Gordon had never tried to do that simple task in the confines of the Gemini cockpit with his suit fully inflated. In the inflated pressure suit, Gordon's arms were so stiff that he did not have the dexterity to snap the gold visor properly into place. Leaning over as far as he could Gordon attempted to get Conrad to help him. But, Conrad's suit was nearly as stiff as the EVA garment that Gordon was wearing. After a good deal of struggling Conrad managed to get the left side of the visor in place but the right side fought the two crewmen until it nearly won.

Eventually Gordon got the right side of the visor snapped in place but in the process cracked the visor. By the time they got the visor on they had eaten up most of their extra "rest" time. Through the exertion and the frustration Gordon had managed to exceed the ability of his suit to cool his body and was sweating quite a bit.

It is worth noting here that the problems encountered during Gemini EVAs had a lot to do with the fact no one knew what to expect during an EVA at the time that the suit design for the G4C system was frozen by NASA. As a result the suits were largely designed with an ability to function while inflated, ruled only by the need to throw switches, fly the spacecraft and stand up in the hatch. There was no consideration for doing complex tasks.

Yet, following Ed White's EVA a huge array of complex tasks had been thrown at the suit and its occupants. Additionally, the Liquid Cooled Garment that was eventually developed and worn under the Apollo suits and proved extremely efficient was not a part of the Gemini suits.

Gordon's Environmental Life Support System, also known as the ELSS had been designed to function in the vacuum of space rather than the pressurized cabin of the Gemini spacecraft. The result was the pack was dumping oxygen into the cabin in the spacecraft which was subsequently automatically dumping that overboard. They couldn't afford to waste this.

Additionally, seeing that his pilot was not in the best shape at that moment Conrad wisely took Gordon off the suit circuit and placed him back on the spacecraft circuit in order to better cool him. It was not a pleasant situation— Dick Gordon found himself heavily taxed and had not even gotten out of the spacecraft yet.

When the time came to open the hatch, Gordon swung the door open and nearly sprung into space in the process. The outgassing of residual oxygen in the environmental system started to carry him away. Conrad grabbed a handy strap on Gordon's suit and kept him near the hatch.

It was now the absence of knowledge handed down from previous crews reached out and hit Gordon.

Exactly like Cernan and Collins before him, Gordon found that the laws of physics weighed heavily on everything and everyone in space, even a top-notch fighter pilot. He needed to make his way to the Agena, yet had no foot restraints and no handholds.

As Cernan stated in his post-flight debriefing all he had to do was wiggle his finger and the results transmitted through his entire body.

Every time Gordon touched the surface of the vehicle an equal yet opposite reaction took hold of his body. His feet went where they weren't intended and his legs when directed one way sent his torso in a completely unwanted direction. He was quite literally fighting himself. His friendly tether was quickly transforming into the snake that had fought Gene Cernan.

Suddenly those simple-to-set-up experiments and that two-minute "nothin' to it" maneuver to fasten the lanyard on to the Agena turned into a wrestling match that lasted 10 times longer.

As Gemini XI came into range of the Hawaii tracking station the flight surgeon's console showed Gordon with a respiration of 36! CAPCOM warned Conrad that his pilot was hyperventilating.

Eventually Gordon's respiration came down to 26, but things were not off to a good start. It was just after 9:30 Eastern time and the EVA was only getting started. After just 18 minutes of struggling with tethers, cameras, and body positioning Gordon told Conrad,

"I have got to rest for a minute… I'm pooped."

After resting for a few minutes with Conrad holding onto his legs Gordon finally managed to gather up enough energy and head for the Agena. He launched himself toward the Agena figuring to get the job done in a snap.

Unfortunately, he overshot the docking adapter and headed off into space. Conrad saw this coming and nabbed the umbilical. Hand over hand, he reeled Gordon back in toward the open hatch.

Now Gordon tried again, using the Gemini's RCS nozzles as handholds he propelled himself to the Agena. Straddling the nose of the Gemini like riding a horse he readied for the task of attaching the lanyard to it and then to the Agena.

"Ride 'em cowboy!" Conrad crowed over the radio.

In training he had been able to accomplish this single exercise in the time it took to do one parabola in the zero G airplane— about 45 seconds. Very quickly, however, Gordon learned that 45 seconds in the airplane was not a good representation of continuous weightlessness in orbit. In the airplane with his feet jammed in the docking adapter he was able to use both hands to make the connection. In space his feet provided no leverage and he had to stabilize himself with one hand. Every time he moved the hand holding the lanyard his entire body started to pull away from the Agena and he had to re-stabilize himself. He struggled for what felt like a half hour but was actually less than 10 minutes and finally secured the lanyard.

Returning to the hatch, Gordon was exhausted. Sweat from his forehead had gotten into his eye and although he could still see out of that eye, the sweat was producing a great deal of stinging discomfort. Gordon's respiration had peaked at 40 and his heart rate had peaked at 162.

Hearing how spent Gordon sounded as well as he appeared and keeping in mind the troubles that other spacewalkers had getting back into the Gemini, Conrad decided to terminate EVA. An EVA that had been scheduled to last 107 minutes had been terminated after just 33 minutes.

(See Image 19, page 166.) Now the difficulties with Gordon's EVA combined with the problems experienced on Cernan's EVA sent something of a shock wave through NASA. Two of the best pilots on the planet Earth, two of the most goal-oriented, get-it-done diehard individuals ever to rip holes in the sky had been beaten down in a matter of minutes by orbital EVA. One of the primary goals Project Gemini was not being met. There was one Gemini flight remaining and NASA had to solve this EVA problem NOW.

Following a sleep period the crew prepared for what was to be a mission highlight, "the high ride." This involved igniting the Agena's PPS engine, burning it for 26 seconds and boosting Gemini XI to an altitude of 853.53 statute miles. The burn began over the Canary Islands station and the crew found themselves in a high-speed elevator away from the Earth.

"Whoop-dee-do!" Conrad exclaimed at cutoff, "look at her go!"

For two full orbits, the crew took photographs and radioed back descriptions from an altitude higher than any human being had ever traveled. From their apogee, Gordon took synoptic terrain and synoptic weather photographs. Additionally they discovered that radiation levels at their altitude were far less than many scientists expected,

"Our dosimeter reads .3 rads per hour up here," Conrad reported.

"Houston," Gordon chimed in, "radiation is revised to .2 rads per hour."

As Gemini XI soared higher above the Earth than mankind had ever before traveled, almost no one in the United States knew it was happening. It was just after two o'clock in the morning Lexington Drive time in Michigan and all of the space-buffs who were growing up with spaceflight were tucked comfortably in their beds.

In the mid 1960s most TV channels across the nation ended their broadcast day at about midnight local time. So even if the networks were carrying a broadcast of the highest flight in history almost none of their affiliate stations would be on the air to carry it.

Following their high flight, the crew of Gemini XI prepared for their scheduled standup EVA. As Dick Gordon was getting ready to open the Hatch, I was getting ready to open the front door and walk to school. It was a grey gloomy mid-September day in Sheridan Park and the network coverage of the upcoming EVA had effectively preempted all of my before-school cartoon favorites.

Yet there was something pretty neat-O about those space-suited guys and their mockup Gemini spaceship on the TV. It occurred to me that it would probably be a lot more fun to stay home all day and watch this rather than sit in Annex 3 and watch Mrs. Shea nap.

Compared to the free-floating EVA, the standup EVA actually was a piece of cake. Gordon was on a short tether and the great portion of his work consisted of photography. Gordon was assigned to photograph assorted star fields while on the night side of the planet using an S-13 ultraviolet astronomical camera. The work was to be done on two different passes.

The first pass would be completed and when the spacecraft neared daylight on the United States West Coast, the astronauts were then supposed to close the hatch and wait until the spacecraft passed over the United States and over the Atlantic Ocean. When they passed back into darkness Gordon would open the hatch again and complete the second pass of photographs.

When Gordon completed the first pass of photographs, CAPCOM advised the crew that the skies over their hometown of Houston were clear. With that in mind, the crew decided that they would leave the hatch open and photograph their hometown.

Following a jovial pass over both Houston and Cape Canaveral the crew of Gemini XI headed toward the next night pass to complete their photography and their standup EVA. Of course, this left the crew with the entire Atlantic Ocean to cross and nothing to do. When the spacecraft came back into radio contact at Tananarive, Conrad had a confession to make. As the crew was out of communications range and passing over the wide open Atlantic Ocean, Conrad was just sitting there strapped into his seat and Gordon was standing in the open hatchway waiting to come up on the night pass. During that time the two astronauts both fell asleep. Zooming through space at 17,500 miles per hour Conrad and Gordon took a nap.

Mrs. Shea would have been proud.

GEMINI XI
LEAVING THE BEST FRIEND THEY EVER HAD

Conrad and Gordon finished up their mission with a never before done series of exercises using their Agena. Conrad freed their spacecraft from the Agena's docking cone and backed away allowing the lanyard that Gordon had tried so hard to connect to un-reel.

Once the lanyard was fully extended, the crew practiced a series of maneuvers designed to allow the Gemini and the Agena to rotate around one another held together by the lanyard.

The objective was to see if two spacecraft could station keep in this manner and save fuel on orbit. Another aspect of the test was to see if, while rotating around one another, the two spacecraft could use centrifugal force to create an artificial gravity. Although the test was not overwhelmingly successful, a small amount of artificial gravity was produced. After three hours in this tied together configuration, the crew jettisoned the Gemini's docking bar and thereby released the Agena.

Considering that the crew of Gemini XI had used far less fuel than expected during the overall mission, controllers in Houston came up with another ad hoc test to throw into the mission. Houston advised the crew to prepare for a coincident-orbit rendezvous with the Agena. Instead of firing his thrusters in retrograde and dropping into a lower orbit thus leaving the Agena behind, Gemini XI would fire their thrusters in posigrade and climb above the Agena. After three quarters of an orbit Conrad would again fire his thrusters, drop in behind the Agena and allow it to catch up. The plan worked just fine and after an hour Conrad performed a braking maneuver and rendezvoused with the Agena.

Having performed perfectly, the Agena now was to be left behind by Gemini XI.

"We made a 3 foot per second retrograde burn and left the best friend we ever had," Conrad said as the crew departed their Agena.

"We were very sorry to see that Agena go," Gordon added, "it was very kind to us."

Having started out as the object of doubt and concern the Gemini Agena Target Vehicle was now the sweetheart of the program.

To top off their mission the crew of Gemini XI now had one final never been done before task to accomplish. They were scheduled to do the program's first all automatic reentry. Conrad was supposed to keep his hands off his controls and let the onboard computer do it all.

Gemini XI splashed down less than a mile from the carrier GUAM and the entire event was carried live on television. It was just after nine o'clock in the morning and I was, of course, stuck in Mrs. Shea's class as she lorded over me, yardstick in hand, enjoying every second as I sweated and stammered through reading aloud some printed nonsense.

Any hesitation on my part would likely draw a swift

swipe in order to re-cage my ever-wandering attention span. Indeed, I should've faked an asthma attack and stayed home to watch the splashdown on TV.

Of course I would even things up with the old battleaxe. Soon another movie would be delivered to Annex 3 and Mrs. Shea would cop another nap whereupon I would draw yet another cartoon of her on a bathroom napkin and use it to crack up the kid next to me.

GEMINI XII
PLUS 10 CENTS FOR SHIPPING AND HANDLING

Following the "Wes can't read," scare of my second grade year and my subsequent flunking of the third grade, my parents were doing the best they could to follow the advice of the reading specialist and provide as much reading material as they could involving any subject that may hold my interest.

In the weeks just prior to Gemini XI, my Mom had spotted a reader reply card in a magazine. Printed on it was a hodgepodge of images and taglines; "Start exploring the world of science around you!" and, "Get this BRAND-NEW model of the three-man Apollo moon ship!" "All this for 10 cents" "build the NEW APOLLO moon ship!" And there was a picture of the Apollo command and service module model that looked like anyone could build it... even me.

It was all from the Science Service- Science Program and for just one dollar per month plus a dime for shipping and handling they would start you off with a model of the Apollo command module and lunar module plus a book "Man In Space" and a nifty wall poster depicting the phases of flight to the Moon.

Calling me over to where she was sitting, Mom ran her finger over the entire ad directing my attention to the various images,

"Do you think you'd be interested in something like that?" She asked with a note of reservation.

"I guess," I half shrugged.

That tiny spark of interest was enough for my parents to invest that dollar a month plus ten cents for shipping and handling. They had no idea that this simple investment would plant the seed that would captivate my interest and focus my perpetually-wandering attention span on the subject of spaceflight. That single dime Mom taped to the little circle that stated "PLACE DIME HERE" and then folded over the first corner and the second corner as instructed would start their son in a direction that he would follow for the rest of his life.

Just days before the flight of Gemini XII my package arrived from the Science Service. Not only was there a model and a poster and a bulletin and a book, but the book had stickers in it! The stickers were of things such as the original seven astronauts holding rocket models, a Convair rocket sled under power, Gemini IV blasting off, fueling an Atlas, Scott Crossfield in a pressure suit being exposed to heat lamps, a radar tracking ship and space monkey Baker. What self-respecting nine-year-old boy could resist such a thing in 1966? I immediately went to work licking and sticking in those images. They drew me in to read and discover what those things were all about.

Then there was the poster— it showed all of the phases of the proposed flight to the moon and had a picture of a LEM, a Gemini spacecraft with a cutaway interior, a Mercury spacecraft with a cutaway interior and for scale, a six-foot astronaut standing among them.

In the background there was the rough and rugged sharp-edged riprap stone and steep cliffs of a lunar landscape. Immediately that poster was stuck to my bedroom wall that previously had been adorned only by Rat Fink hot rod wallpaper. Now I stood back and gazed at the poster completely entranced for my full attention span of about a minute and a half. Eventually I would begin lengthening that envelope a few minutes at a time.

Next there was the model and I had the skill, the glue, the emery board and the single edge razor blade that would bring that Apollo spacecraft to life. Sure it would not be a museum quality thing of beauty, but with the skills of a nine-year-old I actually constructed the thing.

That Command and Service Module (CSM) would eventually fly countless imaginary orbits around the earth and countless trips through cislunar space. A half-century later I still have that command module.

Meanwhile the Lunar Excursion Module (LEM) would make innumerable precision landings on our living room carpet. Within the next three years my slot cars, Hot Wheels and Rat Fink models would be pushed aside in favor of spaceflight. In fact, although I could not imagine it at that moment, even Batman would be overshadowed by the space program.

Yet, just as I was beginning to look at spaceflight from a different perspective, as opposed to Fireball XL5, and turning a newfound interest toward America's current space program, Project Gemini, there was just one Gemini flight remaining.

Just my luck.

GEMINI XII
EVA IN THE MCDONOGH POOL

In the autumn of 1966 astronaut Jim Lovell, the command pilot for Gemini XII had something of a problem on his hands. The way things were going, it began to appear as if there was no real mission for Gemini XII.

Originally the mission had been built around the testing of the Astronaut Maneuvering Unit which had been scrubbed during the mission of Gemini IX-A. Considering the difficulties that astronauts had on the previous three EVA's, upper level management at NASA was now concerned the level of work in deploying and operating the Air Force's AMU made it too risky for the final Gemini flight. The Air Force, however, held the position that a great deal of money had been invested in development of the AMU and it deserved another chance.

But On September 23rd the flight ready AMU that had been mounted in the adapter of the Gemini XII spacecraft was ordered removed. One week later Associate Administrator of the NASA Office of Manned Space Flight George Muller officially notified the United States Air Force the AMU had been deleted from Gemini XII and the core objective of the mission was gone.

All that remain for Gemini XII to do in space were things that had already been done by previous Gemini crews. Rendezvous from below and above, docking, automated landing, long-duration flight and EVA had all been done, yet only one of those tasks remained to be mastered and that was EVA. So somewhat to Lovell's chagrin, Gemini XII would not be breaking any new ground or blazing any trails in space— instead it would be fine tuning the critical task of EVA.

Although Lovell's pilot for the mission, Buzz Aldrin, (yes I know his real name is "Edwin" but he has always considered his first name to be "Buzz" and so it will be here) had been training intensively for about a year to fly the AMU and he was supremely confident in his ability to put it on and operate it, the matter was taken out of his hands. Now he had to focus on training to help solve the riddle of how to do an EVA without exhausting himself.

Some documentaries and poorly researched "histories" of spaceflight imply or sometimes openly state that Buzz Aldrin invented underwater neutral buoyancy training. That is not even close to the truth when it comes to use of neutral buoyancy in simulating spaceflight conditions.

Although Aldrin was the first astronaut to utilize neutral buoyancy training for his upcoming EVA, he did not invent it. His use of neutral buoyancy training did, however, facilitate the bringing together of all the pieces required to make neutral buoyancy a key element in future EVA training.

In fact, as far back as 1964 G. Sam Mattingly and Harry Loats of the Environmental Research Associates along with Otto Trout of NASA's Langley

Research Center were pioneering the concept of neutral buoyancy.

Upon gaining the use of a swimming pool at the McDonogh School in Owings Mills, Maryland by way of permission from Robert Lamborn, the school's Headmaster, they took the Arrowhead version of the Goodrich Mark IV pressure suit and on July 19th, 1964 began simulations of operating an airlock in neutral buoyancy.

Mattingly was the first person in the suit. Less than a month prior, the same trio had attempted a crude version of neutral buoyancy in a pool at Langley. This time in the McDonogh pool, they were conducting a true engineering study. Films of their results were shown to Paul Hill, the Deputy Chief of Structures and Mechanics Division at Langley. Mr. Hill was impressed with the results and a proposal was submitted and resulted in contract NAS1-4059 for the further study and simulation of airlock features and sizes through neutral buoyancy.

Use of neutral buoyancy, however, was given a backseat at NASA.

Since the beginning of astronaut training in the late 1950s zero gravity simulations had always been the territory of aircraft flying parabolic curves. By that method, a C-131 military transport aircraft was able to gain about 15 seconds worth of simulated weightlessness. Later KC-135 jet aircraft were able to gain a little more than 45 seconds of zero gravity.

NASA was a flying agency with its roots in the sky and not underwater. By use of aircraft and frictionless air bed tables, most people in NASA were quite sure that they had all of the problems of weightlessness within their realm of training. There was no reason to go off on a neutral buoyancy tangent; especially late in the Gemini program. Then came Geminis IX, X and XI with unexpected EVA difficulties on all three flights.

Sometime after Gemini IX Dr. Gilruth suspected that the agency's EVA training was on the wrong path. He dispatched Gene Cernan to the McDonogh pool where a Gemini adapter section mockup and its AMU were set up. Cernan's task was to evaluate, compare, and contrast the neutral buoyancy simulation with his EVA and return results.

What Cernan discovered was that the neutral buoyancy simulation was very much like what he had experienced in orbit. At that time there had been some quiet questions around NASA as to whether or not Cernan's difficulties on the Gemini IX E V A were simply his fault. Mattingly later stated that when Cernan finished his simulation he expressed a great deal of relief— his problems with the AMU were clearly not his own.

On the day Gemini XI launched, Buzz Aldrin was actually in the McDonogh pool training for Gemini XII. Mattingly recalled that they rolled the TV set into the room where the pool was located and Aldrin, clad in his pressure suit, stood in the four-foot deep end of the pool and watched the launch before beginning that day's simulation.

(See Image 20, page 167.) Both Aldrin and his back-up Cernan trained intensely in the pool. Mattingly also recalled that films of previous simulations had been sent to Houston prior to Gemini XI but no sources indicate that those had ever been included in

Dick Gordon's training.

As the launch day for Gemini XII approached, Aldrin's training had been focused upon handholds, foot restraints, tools and body tethers. Aldrin himself was somewhat frustrated with the lack of challenge not only in the training but in the objectives of the mission. He felt as if he was being sent into space to do little more than what your neighborhood handyman would do. Oddly, following the mission, history would depict Aldrin's efforts and EVAs in a very different light.

GEMINI XII
WHAT… ME WORRY?

Just four days after the launch of Gemini XI the Gemini XII Titan II launch vehicle was moved to Complex 19 and erected on the pad.

Four days later, on September 23rd, Gemini spacecraft 12 was moved to Complex 19 and stacked aboard the launch vehicle.

Three days after that Atlas 5307, the target launch vehicle for Gemini XII was erected at Complex 14.

Previous failures of both Atlas' and Agena's had left the Gemini program short of both by exactly one.

The Agena shortage was relatively easy to fix; vehicle 5001 that had been used in fit trials had been sent back to the manufacturer and Lockheed had refitted it for use as a flight article. It was now compatible with Gemini spacecraft 12.

The Atlas booster, however, was a much larger problem. General Dynamics Corporation, who made the Atlas, did not stockpile boosters hoping that one day someone would come along and buy one. By 1966, all of their Atlas boosters for that year had been committed to other launches. NASA's Gemini Program Office had to go looking within the agency to find an Atlas booster that could be redirected for the Gemini XII mission.

As luck would have it, Langley's Lunar Orbiter II had run into problems. The vehicle intended for that mission was actually Spacecraft 5 but a NASA review of equipment showed that a thermal door had failed to open during vacuum testing and the spacecraft was sent back to Boeing for corrective action. The delay caused a shuffling in boosters and subsequently freed up Atlas 5307, which the Gemini Program Office soon commandeered for Gemini XII's Agena.

Interestingly, Lunar Orbiter II still ended up launching five days earlier than Gemini XII. The Gemini Program Office had negotiated a swap with an Atlas booster from Vandenberg to support the Lunar Orbiter. That booster, however, was a new version of the Atlas that had previously been un-flown and NASA officials did not want to risk their last Agena on an un-flown vehicle. So they kept Atlas 5307 and gave the new vehicle to the Lunar Orbiter program.

Of course, it seemed you couldn't get a Gemini flight off the ground without a launch delay of some sort; Gemini XII was no different. No sooner had the countdown begun for the November 9th scheduled launch when a problem in a power supply in a secondary autopilot occurred. The launch was scrubbed before the crew was even suited up and it was rescheduled for November 11th.

On launch day Lovell and Aldrin exited the suit-up quarters; Lovell wearing a paper sign in the middle of his back that read "THE" and Aldrin wearing a similar sign that read "END." The crew boarded the transfer van and made the final trip along the Barton FREEway.

Aldrin was also given a paper face taped to the top

of his helmet that when flipped down would show a drawing of "MAD" magazine's Alfred E. Newman and the catchphrase "What, me worry?"

Pad leader Guenter Wendt and his ever-reliable McDonnell white room team met the crew. Lovell handed Wendt a 4-foot long check made out for 1 million Deutschmarks and said to be for "severance pay." That was highly appropriate, because 3.7 seconds after the countdown reached zero and Gemini XII left Cape Canaveral so would the jobs of every contract employee who had been working there on Project Gemini.

NASA employees were integrated into Apollo, but McDonnell employees were no longer needed. The contractors for Apollo were North American Aviation, Boeing and Grumman— so if you were an employee of the Martin Company or McDonnell, as the phases of the Gemini XII mission ended so did your employment.

Some Martin employees were transferred to the Titan III program and managed to stay at the Cape, but most of the McDonnell employees had seen their last day in the space program.

Wendt, however, was seen by the astronauts as a non-replaceable item. Six months after the Apollo 1 fire he was hired by North American and returned to his Pad Leader position with the full and absolute authority that he had been given for both Mercury and Gemini. He would serve through Apollo and into the Shuttle program before retiring from Spaceflight in 1989.

Launch time for the GT-12 Agena was set for 2:49 PM Sheridan Park time. Atlas 5307 boosted GATV

5001 flawlessly into orbit precisely on time.

Launch of the astronauts was scheduled for 3:28 PM Lexington Drive time which, for a change, was perfect for those of us in the Eastern time zone who were growing up with spaceflight. Nelle Haley dismissed at 3:15 and that gave me plenty of time to escape Annex 3, slog home, take off my jacket and boots and park in front of the TV set.

Considering it was a balmy 41 degrees outside, my Mom would surely kick me out of the house and tell me to go play soon enough. Today, however, Mom was quite content for me to park in front of the TV set and watch the news coverage of the GT-12 launch.

Granted, I didn't know what the heck the newsmen were talking about, but I knew that when that countdown clock reached zero I was gonna see a big rocket takeoff.

GT-12 did not disappoint me. The booster lifted off cleanly and then punched through a layer of clouds and headed toward space. By the time the second stage animation concluded, my interest was beginning to wane and Mom booted me outside.

As Gemini XII successfully entered orbit, Launch Complex 19 and Launch Complex 16 were being vacated.

The Titan vehicle erector at Pad 19 would never again be raised. Instead it would be left to rust away in the salt breezes from the nearby Atlantic Ocean. Because of the hypergolic propellant's toxicity the entire area of Launch Complex 19 would be declared a hazard and posted as "off-limits" to all personnel. It is a restriction that remains to this day.

The Barton FREEway would be left to the Florida

scrub to grow over it.

In the early 1970s, the umbilical tower at Pad 19 would be torn down and sold as scrap metal. Perhaps the only thing left to us from Complex 19 is the giant aluminum White Room that serviced the Gemini spacecrafts. In the early 2000's that white room was removed from the pad by volunteers of the Air Force Space and Missile Museum at Cape Canaveral Air Force Station. It was restored and placed on the Museum grounds at launch complex 26.

While Lovell and Aldrin were rendezvousing with their Agena they lost radar lock and found that their computer was no longer operating. Aldrin, who was known around NASA as "Dr. Rendezvous" took over with his charts and his sextant and calculated his way to a successful rendezvous and docking.

Plans to fire the Agena PPS in place Gemini XII into a higher orbit had to be canceled as a previous firing of the Agena's engine prior to docking had shown a hic-up in thrust chamber pressure. Still cautious about the capabilities of the Agena, NASA flight directors decided not to re-fire the engine. In the subsequent re-planning it was decided to alter Gemini XII's flight plan and orbit in order to send the crew on a different sort of rendezvous— this time with an eclipse of the sun.

As a matter of fact, the solar eclipse rendezvous had been placed into Gemini XII's flight plan in its early phases. But then was removed in favor of other experiments. So, Lovell and Aldrin were highly in favor of the change that would allow them to seek out the eclipse. The only problem was mission control decided to radio all of the required information up to them right while they were trying to eat. By using the Agena's secondary propulsion system the crew was able to conduct a series of phasing maneuvers that would realign Gemini XII with the path of the eclipse.

On the second day of the mission, Aldrin easily knocked off the first stand up EVA. A total of two hours and 29 minutes were spent performing experiments, taking photographs and using periods of rest. Following that, Aldrin easily closed and locked the hatch. Gone were the days of sticky hatches that refused to close and astronauts exhausting themselves trying simply to get back into the spacecraft.

The following day, however, was set to be the big test for EVA— a free-floating umbilical EVA. Essentially it was the exact same type of EVA that had exhausted Gordon on the previous mission.

The exception was that this time Aldrin was equipped with a vast array of handholds, lanyards, short tethers and, most importantly, the "golden slippers."

Although as spaceflight advanced, G4C spacesuits would soon become obsolete— as would zip guns and bulky umbilicals— the golden slippers would remain. To this day foot restraints similar to the golden slippers are still in use by astronauts.

The golden slippers were very simply a rubberized sort of one-third house slipper. The object was the astronaut would take his feet and point them toe-in when placing them into the slippers. The natural position for the feet is to point slightly toe out, so when the astronaut relaxed their ankles their toes would naturally snap into the slippers. That was enough restraint to keep a working astronaut in position in zero gravity.

In the neutral buoyancy tank Aldrin found that the golden slippers held him in position so well that he could lean forward or backward 90° and sideways more than 45° and his feet would remain within the slippers. On Aldrin's tethered EVA, the slippers worked exactly the way they did in the neutral buoyancy tank. Additionally all of the handholds, short tethers, and other positioning aids worked equally well. Aldrin was able to leave the Gemini, move easily to the Agena, and connect a line in a matter of seconds. It was this task that had exhausted Gordon on the previous flight.

Aldrin then moved to the spacecraft adapter, inserted himself in the golden slippers, and proceeded to work at what they called the "busy box." In this box he cut metal, drilled holes, turned nuts and bolts and did assorted other tasks that would have easily exhausted previous astronauts.

It was not that Aldrin was some sort of Superman or that he was in better physical condition than anyone else that allowed him to perform these tasks with such ease— it was the new methodology. There was a new way of thinking about harnesses for restraints and handholds along with the experience and the neutral buoyancy tank that allowed Aldrin to successfully complete his EVA's.

For two hours and eight minutes, Aldrin performed all the chores that had exhausted everyone else, then returned to the spacecraft without breaking a sweat. The following day he performed a second stand up EVA with an equal amount of ease. NASA was now sure that the monster of EVA had been tamed.

Following another gravity gradient experiment

with the Agena tethered to the Gemini and a few other in-cockpit experiments, it was time for Gemini XII to return to earth.

Meanwhile, back on Earth the tectonic motion of Apollo was taking place. NASA had just announced a contract with International Latex Corporation for $10 million to produce the Apollo spacesuit. Consisting of the undergarment, a liquid-cooled constant wear garment, pressure bladder assembly, plus thermo-micrometeoroid protective outer garment, it was what would become the Apollo spacesuit.

Coinciding with that, a $20 million contract was settled with Hamilton Standard Division of United Aircraft Corporation ordering them to continue development and manufacture of the Portable Life Support System (PLSS) backpack.

Originally Hamilton Standard had been the prime contractor for the suit with International Latex as the sub-contractor, but the two did not play well together and NASA cancelled the deal. Now the two were on their own and the result took us onto the lunar surface.

Also, as Gemini XII prepared to come home Saturn IB AS-204 was stacked on Pad 34 at the Cape and the Apollo 1 command module was attempting to get its bugs cleaned out at the Operations and Checkout facility.

Designers were still having issues with the electrical system as well as the crew couches. Recent tests had showed that in a normal landing impact the PLSS backpacks, which were stored below the seats, would impact the astronaut's helmets and make about a four inch dent— and that was just one of countless problems that needed to be solved.

The Block I Service Propulsion System engine, however, successfully completed its first altitude chamber firing.

Saturn IB equipment was being readied for the planned Apollo Applications Program and planners were looking at the need for as many as two dozen Saturn IB boosters to support that program. And, at White Sands Missile Range the final Little Joe II test launch of the Apollo Launch Escape System was being readied. America's space program was moving ahead at break-neck speed.

When Lovell and Aldrin stepped onto the deck of the aircraft carrier WASP they were not only closing out the Gemini program but they were also about to step into the whirlwind of the Apollo program.

Of course, I was being held captive in Annex 3 at 2:49 in the afternoon of November 15th, 1966 when the crew of Gemini XII stepped onto the red carpet aboard the carrier. So once again I missed television coverage of that event.

But, Mrs. Shea's afternoon nap and movie still had about 10 minutes to go— so I wasn't in bad shape.

Frankly, I don't think the reality that the Gemini program was actually coming to an end had really sunk in with most kids my age who were growing up with spaceflight. After all, there had been another Gemini launching every two months for the past two years, so in a kid's mind there would always be another one coming without regard to the fact the people in the news media had been telling us repeatedly the Gemini XII was the end.

Model shelves in the toy departments of every store still carried Gemini models and model maker Revell had actually devised an amazing publicity stunt that was all set to be released upon the growing up with spaceflight public in February 1967.

The company was going to have a sweepstakes wherein some lucky kid would win a FULL SCALE Gemini spacecraft. Months after the final Gemini flight a 13-year-old boy by the name of Robbie Hanshew, from Portland, Oregon was the big winner. He subsequently donated the full-sized Gemini spacecraft to the Oregon Museum of Science and Industry where it resides to this day. The Gemini was reportedly so large and heavy that it had to be delivered by railcar. Had I been lucky enough to win the Gemini that would've worked out perfectly for me— the tracks of the Grand Trunk Railroad ran right through my backyard in 1967. (See Image 21, page 167.)

(MEMORIES OF OTHER FOLKS) Richard; USA: "My father started working for the McDonnell Aircraft Corporation ("Mac") on the Gemini program from 1962 to the completion of the program helping to assemble, test, and launch various Gemini spacecraft including the Gemini IV, IX, X, XI, and XII spacecraft.

One day while working on the Gemini IV spacecraft (he was the Operations Engineer), my father was informed that the astronauts scheduled for the flight were coming for an inspection tour and the work he was doing had to be finished before they arrived. Feeling the pressure my father started to complain about rushing the work just to have a photo opportunity with the astronauts.

As he was working, an unknown person came up to where my father was, started to watch what he was doing, and then began talking to my father. My father continued to complain about the rush and said that he did not care if he completed the work by the time the astronauts arrived. As he was more concerned about the quality of the work and he was going to ensure that the work done was correct with no errors.

During this time, the unknown person asked if there was anything he could do to help my father. While they were talking, the unknown person actually did help my father.

After a while the unknown person said that he had to leave. My father was able to complete the work before the scheduled time and of course done correctly.

At the scheduled time, the inspection group arrived and my father immediately became embarrassed when he realized that the unknown person was in fact one of the two astronauts.

After the tour, but before he left, the astronaut walked up to my father and shook his hand to thank him. He told my father that he appreciated my father's attitude and he was glad to hear how my father was very concerned with the safety of the spacecraft and crew to ensure that the work performed was correct, as his life depended upon that spacecraft.

I do not remember which astronaut it was but according to my younger brother, it was Ed White. It turned out that Ed White was so impressed with my father that after the mission, he gave my father a Gemini IV mission patch that flew in space with the astronauts. My older brother has the patch given to my father.

James McDonnell the founder, then president, and CEO of Mac was a strong supporter of the Boy

Scouts of America and was actively involved with the St. Louis organization, which is where my father met and got to know him.

As the scoutmaster for our troop 803, my father turned in a vacation request for a week off so he could attend the yearly scout jamboree but the request was denied due to a heavy workload.

Just by coincidence, James McDonnell came by my father's area and while he was there, he started talking to my father. During the conversation, my father mentioned in passing that he would not be able to attend the jamboree due to the work schedule for the upcoming Gemini launches. When done talking, James McDonnell left and my father went back to work.

The next day my father was asked to see his boss and when he went in to talk to him, he found out that his vacation request was approved. It turned out that James McDonnell sent a directive down to my father's boss to have him change my father's work schedule and approve his vacation.

One day my father was working on a Gemini spacecraft when all of a sudden, a T-38 aircraft carrying the prime crew for the Gemini IX mission came crashing through the roof just feet from where my father was working.

Fortunately for the Mac personnel, the T-38 bounced off the roof and did not crash all the way through it, so only a few people were injured on the ground, and my father escaped injury. Unfortunately, for Elliot See Jr. and Charles Bassett II, after bouncing off the roof, they then crashed into an adjacent courtyard and were killed on impact.

The backup crew of Thomas Stafford and Eugene Cernan flying in another T-38 just behind the primary crew landed safely. All four of the astronauts were flying into St. Louis for two weeks of training in a Gemini simulator. This accident shocked everyone in the area and my father never wanted to talk about any of the details from that day.

An example of how important the space race was to the nation during that time in history, the Mac Gemini launch crew from St. Louis was scheduled to be at the KSC for over four months for the launches of Gemini IX through XII.

A decision was made by either the government or Mac that for the morale of the launch crew, the families of those involved would accompany the crew on an all-expense paid trip to Florida during the summer of 1966.

While the Mac employees had to work, we had a paid vacation in Florida including going out to restaurants many times. As a family, we spent only 14 dollars of our own money the entire two months we were in Florida.

The only problem we had in Florida was we felt that the water had a bad taste so we had to drink Tang, Kool-aid, and milk and being kids we did not mind.

While in Florida, we were able to film one launch of a booster vehicle carrying one or more satellites but unfortunately, just after we ran out of film the booster vehicle exploded destroying the satellites. It was quite a sight to watch all of the burning fragments fall back to earth.

This was the first of two launch failures that I have watched in person with the second being many years later when in April 1986 the 20th launch attempt of the HEXAGON photoreconnaissance satellite program exploded eight seconds into the launch. We were not allowed to go swimming in the ocean for a couple of days after the launch explosion due to the fuel contaminating the ocean. Every time we watched the movie of the launch, we relived the frustration of not filming the actual explosion.

We were lucky enough to have heard about a night test launch of a Polaris Submarine Launched Fleet Ballistic Missile (SLBM) from a submerged submarine off the coast of the KSC. It was an amazing sight when out of the dark night and from the middle of the ocean, the solid rocket motor of the Polaris missile ignited and arched across the sky as it climbed downrange.

The ironic thing about our trip to Florida was because my father had to work during the launch of Gemini X we were not able to get on base to watch the launch in person so we had to watch the launch on TV like most of the nation.

During one of my father's days off, my family went on a self-guided tour of the KSC and as we were driving next to one of the "creepy crawler's" Crawlerways, we decided to stop and take some of the rocks from the Crawlerway as a souvenir of our trip.

Shortly after we scooped up some of the rocks into a small jar we had in the car a security vehicle pulled up and asked us what we were doing out of our car. We explained to him that we were taking a closer look at the Crawlerway. He responded by telling us that we were not allowed out of our car so we got back in our car and continued our tour.

When we took a close look at the rocks that we had

in the jar we were surprised to find that the "creepy crawler" had crushed some of the rock into a powder as fine as mascara. We still have a copy of the drive thru tour booklet for NASA's KSC and the Cape Kennedy Air Force Station giving a brief description of the various highlights of the tour.

Even though NASA called it the crawler-transporter, it had a nickname of "creepy crawler."

To top off our trip, my family including our dog (except for my father) had to leave early, so we were flown back to St. Louis at either the government's or Mac's expense.

Because of my father's job, we were living the space race as a family and we still have several mementos that have flown in space from the Gemini program. Fortunately, for me my father's involvement with space programs did not end with Gemini so my childhood excitement with space never ended.

My father was very proud of the Gemini program and of his involvement with the top-secret/BYEMAN DORIAN (MOL/Gemini B) and HEXAGON space programs. His name is etched in the Space Walk of Fame Gemini II monument in Titusville, Florida under the McDonnell Aircraft section."

PROJECT GEMINI
SO, WHERE'D YOU GET SOME OF THIS STUFF?

It occurs to me that some of you who may have read this book may perhaps be Gemini-buffs. As such, you may have found some bits of obscure information in here that you have not seen before or have seen documented differently. It is only fair, since I hate footnotes and end notes and all of that proper rot, that I should give some insight as to my sources and methods.

First off, the sources: go to the section of this book titled "sources." They are all listed there.

My methods are the same that I sharpened while researching and writing the historical narratives in my 17 (to date) books about Great Lakes shipwrecks. Sure, I do plenty of maritime history— but researching history is researching history and thus the same methods can be applied to spaceflight. What I do is dig into the facts until I have as much as I can get… then I dig a lot deeper. I look in places that other people do not think to look. I pick up a sliver here and connect it to an obscure sliver from another source and then find the documented connection. No sliver is too small and every sliver has to be questioned. I take a sliver and go off on a tangent to learn more about it.

For example, the exact time in the countdown

when the Titan II gimbal test takes place. Never knew that eh? How'd I come up with it? The fact is— no one had ever published that information, so now it was up to me to sleuth it out— exactly what I LOVE to do.

First, I went to the internet and at the website "archive.org/details/geminiaudiocollection" all of the Gemini voice loops can be found. But, they are filled with static and cross-talk and each recording runs for more than three hours. Now I had to find a way to sort it out.

Using my Sony Vegas video software I opened the audio tracks that I had downloaded (Almost every video editing package will also edit just audio). There I discovered that each audio file was actually TWO files- one on the left channel and one on the right, recorded over one another.

By singling them and listening one channel at a time I was able to get some very clear historic audio. This turned into a very handy tool used throughout the book. (The only problem being that I had to sit and listen to audio for hours and hours.)

In the case of the pre-valves and gimbal checks, all I needed was the countdown audio for one mission. Keep in mind that the exact same countdown sequence was used for these items on every mission. Thus, I got what I wanted during the countdown for Gemini VII. The TC gave a time hack and using a digital timer I simply marked when he called each event. Of course, that little hunt took, believe it or not, almost three full days for just those three little tiny facts.

Finding something such as the time that it took to raise the erector after the Gemini VI shutdown was a good deal more simple. I went to my collection of historic spaceflight news coverage. On the CBS broadcast, Walter Cronkite had the launch pad on a monitor over his shoulder. There was the erector in clear view and the moment that it began to move was easily seen. That video went on continuously and eventually switched to the white room camera and you can see the actual shaking stop when the erector mounts on its legs. I put a stopwatch on it and bingo, I had the time.

I hate to use Wikipedia as a primary source since there are so many ways for odd-balls to add invalid information. However, one thing that the site is good for is fast-tracking you to a valid source. In one case I wanted to know more about neutral buoyancy training. I started on Wikipedia but went immediately to their listed sources for their article. There I found a piece written in "The Space Review" by the guy who invented the process!

Often I use weather as a scene-setter for my narratives. Many of my readers have told me that when they read the story they felt as if they were there.

Of course the weather has to be correct if it is to be used in a historical narrative. It so happens that along with being a space-buff, I am also a weather-buff and can read a weather chart like most people read a newspaper. I got an "A" in advanced meteorology in college because more than half the battle was reading the maps correctly. NOAA has a Central Library on-line where you can pull up US daily weather maps going back as far as 1871. Thus, for someone who can read the maps it is quite easy to take a date for a given event in a given location and accurately describe the weather. That is how I did it along with my own memory of those days way back when.

Nearly every single person who was hands-on the manned space program in any major way, especially astronauts, has contributed to the JSC Oral History project. I highly suggest going there and reading— a LOT. I gleaned so many interesting details about the missions by way of the Oral History project all of which came from the mouths of the people who were there. It helped this series of books tremendously.

Additionally, considering that I was very young during Gemini and not yet highly interested in the program, plus locked up in Annex 3 most of the time, I needed something that would add a more personal element of actually growing up with Gemini. Thus, I went onto social media, NASAspaceflight.com and the Space Hipsters page on Facebook and asked for the memories of other folks who were a bit older and more interested than I.

I got a lot of replies and made the deal that I post would everyone by first name only (or whatever name they wanted to be listed under) and their memory as they had sent it to me plus their city and state. Then I worked that into the text where it seemed to best fit.

Some people supplied stories and did not want their city or state mentioned so it just says "USA" in the book. Others apparently decided they did not want to be in the book and so they did not provide permission for me to publish and they were left out. Overall, I like those little memories and I think it adds to the unique flavor of a spaceflight narrative.

Finally, if you can find a copy of the book "Gemini!" by Gus Grissom— BUY IT, no matter what the cost. That book was actually written by Gus during the

Gemini program. When he died his editor and his wife saw to it that the final product was published. I got my copy back in 1974 from my high school library— they were selling off "little-used books" and mine cost me something like 50 cents. It is a treasure in my collection.

So, what you have read in this book was not someone's opinion of what went on or what happened— this is a true story and a heavily researched narrative. If this is your first of my series of Growing up with Spaceflight, I can assure you that the other five are written with the same focused research.

SOURCES

"Gemini!" Gus Grissom

"Titan II" David K. Stumpf

"Gemini" David J. Shayler

"How NASA Learned to Fly in Space" David M. Harland

"The History of Manned Spaceflight" Dr. David Baker

"Two into the Blue" Robert L. Adcock

"Forever Young" John W. Young and James R. Hansen

"Last Man on the Moon" Eugene Cernan and Don Davis

"Carrying the Fire" Michael Collins

"We Have Capture" Tom Stafford with Michael Cassutt

"Two Sides of the Moon" David R. Scott, Alexi Leonov

"Failure is Not an Option" Gene Kranz

"US Spacesuits" Kenneth S. Thomas and Harold J. McMann

"Schirra's Space" Wally Schirra with Richard N, Billings

"The Space Shuttle Decision, 1965-1972" Vol. 1, T.A. Heppenheimer

"Project Apollo, Mission to the Moon," Charles Coombs

"What The Moon Astronauts Will Do All Day," Robert W. Hill

"Gemini and Apollo," Gardner Soule

"Moon Lander: How We Developed The Apollo Lunar Module," Thomas J. Kelly

"Return to Earth," Colonel Edwin E. "Buzz" Aldrin, Jr with Wayne Warga

"We Reach the Moon," John Noble Wilford

"Tracking Apollo to the Moon," Hamish Lindsay

"The Unbroken Chain," Guenter Wendt & Russell Still

"Live From Cape Canaveral," Jay Barbree

"Titan I Missile Weapons System," USAF, republished by Periscope Film LLC

"NASA Project Gemini Familiarization Manual," Periscope Film LLC reproduction

"Gemini 3 - The NASA Mission Reports," Robert Godwin

"Gemini 4 - The NASA Mission Reports," Robert Godwin

"Gemini 6 - The NASA Mission Reports," Robert Godwin

"Gemini 7 - The NASA Mission Reports," Robert Godwin

"Gemini 12 - The NASA Mission Reports," Robert Godwin

"On the Shoulders of Titans," Barton C. Hacker and James M. Grimwood

"Flying Without Wings," Milton O. Thompson and Curtis Peebles

"Aiming at Targets," Robert C. Seamans Jr.

"Flight: My Life in Mission Control," Chris Kraft

"Go for Launch," Powell

"Project Gemini Technology and Operations" NASA, SP-4002

Technical Facilities Resume, Rendezvous Docking Simulator, NASA, Langley Research Center, July 1, 1966

NASA Tracking Ship Navigation Systems, Goddard Spaceflight Center, TM-X-72218, December, 1975

Development and Qualification of Gemini Escape

System, NASA HQ, TN D-4031, June, 1967

Gemini Land Development Landing System

Program Volume I - Full-Scale Investigations NASA HQ, TN D-3869, March, 1967

Gemini Land Development Landing System

Program Volume II - Full-Scale Investigations NASA HQ, TN D-3870, March, 1967

Gemini Land Development Landing System

Program Volume III - Full-Scale Investigations NASA HQ, TN D-3871, March, 1967

RELEASE NO: 65-81, Gemini Titan 3, NASA, March 17, 1965

GT-3 Summary Flight Plan, NASA, (undated)

GT-3 Air-to-Ground Transmissions transcript, NASA, (undated)

Gemini Program Mission Report, Gemini 3, Supplemental Report 5 Air-To-Ground Voice Transcription, NASA, MSC-G-R-65-2, June 23, 1965

Composite Air-To-Ground and Onboard Voice Tape Transcription of the GT-3 Mission, NASA, MSC, April 1965

A Review of Spacecraft Waste-Management Systems, NASA, MSC, August 1969

RELEASE NO: 65-158, Gemini 4, NASA, May 12, 1965

Gemini 4 Flight, Fact sheet 291-B 1965, NASA, MSC, N66-90215

Gemini 4 Report Paration and Activities Plan, and, Data Requirements and Priorities, NASA, MSC, TM-Xr-72327, June 1, 1965

Composite Air-To-Ground and Onboard Voice Tape Transcription of the Gemini 4 Mission, NASA Program Gemini Working Paper No. 5035, MSC, August 31, 1965

Gemini 4 Mission Commentary Transcript, NASA, PAO (undated)

RELEASE NO: 65-262, Gemini Titan 5, NASA, August 12, 1965

Food Consumption on the Gemini IV, V, and VII Missions, NASA, TM X-58010, MSC, October 1967

Manned Space- Flight Experiments Interim Report Gemini V Mission, Auditorium of the Federal Office Building-6 Washington, D. C., NASA, January 6, 1966

Gemini V Mission Data, Prepared by McDonnell, Lambert Field St. Louis, Missouri, (undated)

NASA Program Gemini Working Paper No. 5040, Gemini V Air-To-Ground Transcription, MSC, October 5, 1965

RELEASE NO: 65-362, Gemini Titan 7/6, NASA, November 29, 1965

Gemini VI Voice Communications Air-to-Ground, Ground-to-Air and On-Board Transcription, NASA, Control No. C-115269 (undated)

Mission Operation Report Gemini Flight Number Six, NASA, M-913-65-06, October 14, 1965

Gemini VII Mission PAO Commentary, NASA, (undated)

Gemini VII Voice Communications Air-to-Ground and Ground-To-Air and On-Board Transcription Volume I of III

Gemini VII Voice Communications Air-to-Ground and Ground-To-Air and On-Board Transcription Volume II of III

Gemini VII Voice Communications Air-to-Ground and Ground-To-Air and On-Board Transcription Volume III of III

RELEASE NO: 66-52, Gemini 8, NASA, March 15, 1966

RELEASE NO: 66-163, Gemini 8 Experiments, NASA, June 28, 1966

Gemini VIII Mission PAO Commentary, NASA, (undated)

Gemini VIII Voice Communications Air-to-Ground, Ground-to-Air and On-Board Transcription, NASA, MAC Control No. C-115471 (undated)

RELEASE NO: 66-97, Gemini 9, NASA, May 10,

1966

Manned Space-Flight Experiments Interim Report Gemini IX Mission, NASA, N67-16021, Washington, DC, November 1, 1966

Gemini IX Voice Communications Air-to-Ground, Ground-to-Air and On-Board Transcription, NASA, MAC Control No. C-115803 (undated)

Gemini IX Mission PAO Commentary, NASA, (undated)

RELEASE NO: 66-97, Gemini 10, NASA, July 15, 1966

Gemini X (Final) Flight Plan, NASA Flight Crew Support Division, MSC, June 22, 1966

Gemini X Mission PAO Commentary, NASA, (undated)

Gemini X Voice Communications Air-to-Ground, Ground-to-Air and On-Board Transcription, NASA, MAC Control No. C-115883 (undated)

Manned Space-Flight Experiments Interim Report Gemini X Mission, NASA, MSC, TA-R-67-1, July 18-21, 1966

Gemini Program Mission Report Gemini X, NASA, MSC G-R-66-7, August, 1966

RELEASE NO: 66-226, Gemini 11, NASA, September 2, 1966

RELEASE NO: 66-226, Gemini 11, NASA,

September 2, 1966

Gemini XI Mission PAO Commentary, NASA, (undated)

Gemini XI Voice Communications Air-to-Ground, Ground-to-Air and On-Board Transcription, NASA, MAC Control No. C-115958 (undated)

RELEASE NO: 66-272, Gemini 12, NASA, November 3, 1966

Gemini XI Mission PAO Commentary, NASA, (undated)

Gemini XII Voice Communications Air-to-Ground, Ground-to-Air and On-Board Transcription, NASA, MAC Control No. C-116106 (undated)

The Space Review, Monday, February 4, 2013 "A Personal History of Underwater Neutral Buoyancy Simulation," by G. Sam Mattingly, with John B. Charles

Peter W. Merlin, "Paresev Flight log," NASA Dryden Flight Research Center History Office, September 1998

James Mcbarron II, Retired, NASA JSC, U.S. Spacesuit Knowledge Capture Series, "Spacesuit Development and Qualification for Project Gemini" December 4, 2012

NASA Oral History Transcript, James A. McDivitt, Interviewed by Doug Ward, Elk Lake, Michigan, 29 June 1999

NASA Johnson Space Center Oral History Project, Oral History Transcript, Elmer E. Barton, Interviewed by Rebecca Wright, Houston, Texas – 12 April 2000

NASA Johnson Space Center Oral History Transcript, James A. McDivitt, Interviewed by Doug Ward, Elk Lake, Michigan – 29 June 1999

NASA Johnson Space Center Oral History Transcript, John D. Hodge, Interviewed by Rebecca Wright, Great Falls, Virginia -18 April 1999

NASA Johnson Space Center Oral History Transcript, Frank Borman, Interviewed by Catherine Harwood Las Cruces, New Mexico – 13 April 1999

NASA Johnson Space Center Oral History Transcript, James A. Lovell, Jr., Interviewed by Ron Stone, Houston, Texas – 25 May 1999

NASA Johnson Space Center Oral History Project, Oral History Transcript, Fred Gifford, Interviewed by Ramona Wright, Houston, Texas – 12 April 1998

NASA Johnson Space Center Oral History Transcript, John F. Yardley, Interviewed by Summer Chick Bergen, St. Louis, Missouri – 29 June 1998

NASA Johnson Space Center Oral History Transcript, Richard F. Gordon, Jr., Interviewed by Michelle Kelly, Houston, Texas – 17 October 1997

NASA Johnson Space Center Oral History Transcript, Richard F. Gordon, Jr. Interviewed by Catherine Harwood, Houston, Texas – 16 June 1999

Space World Magazine, "The Two Man Mercury Capsule" Willy Ley, March, 1962

Space World Magazine, "America in Space" Martin Caidin, May, 1962

Space World magazine, "Spacesuits" James V. Correale and Walter W. Guy, September-October 1963

Space World Magazine, "Cryogenics…Products for the Realm of Supercold" November-December 1963

Space World Magazine, "Mercury Experience Applied" Jerome B. Hammack and Walter J. Kapryan, February, 1964

Space World Magazine, "Gemini Launch Schedule" June, 1964

Space World Magazine, "Democracy in Space" Edward H. White, September, 1964

Space World Magazine, "Factors Effecting Food Development For Space" John W. Young (no foolin'), September, 1964

Space World Magazine, "M.O.R.L." November, 1964

Space World Magazine, "Basic Systems For Many Missions" B.H. Rowlett, December, 1964

Space World Magazine, "Underwater Space" December, 1964

Space World Magazine, "Gemini Space Suit", January, 1965

Space World Magazine, "Voskhod Over the Planet", February, 1965

Space World Magazine, "Frictionless Simulator", March, 1965

Space World Magazine, "Russian Report; Man Opens the Door to Outer Space" Boris Adamovich, July, 1965

Space World Magazine, "The World Where Nothing Falls Without A Push" Dr. Robert B. Voas, May, 1965

Space World Magazine, "In the News" (Gemini Spacecraft mirrors) May, 1965

Space World Magazine, "Gemini-Titan 3", June, 1965

Space World Magazine, "Declassified Documents Offer a New Perspective on Yuri Gagarin's Flight," Asif Siddiqi, June, 1965

Space World Magazine, "Gemini 4; Four Days in Space" August, 1965

Space World Magazine, "Shatterproof Helmet" August, 1965

Space World Magazine, "Gemini 5" November,

1965

Space World Magazine, "Gemini 7/6" January, 1966

Space World Magazine, "Atlas and Agena" January, 1966

Space World Magazine, "Agena Engine" March, 1966

Space World Magazine, "Gemini 8" May, 1966

Space World Magazine, "Gemini 9" October, 1966

Space World Magazine, "Gemini 11; Record Breaker" December, 1966

Space World Magazine, "Gemini 12" January, 1967

Boy's Life Magazine, February, 1968

Space news Roundup, "Accident Board Reports Findings in See-Bassett Crash" JSC Newsletter, June 10, 1966

Langley Research Center film, NASA, Serial L-849, "A Water Immersion Technique For The Study Of Mobility of A Pressure-Suited Subject Under Balanced-Gravity Conditions," by Otto F Trout, Harry L. Loats, G. Samuel Mattingly, 1964

Bruce K. Byers, "DESTINATION MOON: A History of the Lunar Orbiter Program", NASA, April 1977

Alfred Robert Hogan, Master of Arts, Thesis; "Televising The Space Age: A Descriptive

Chronology of CBS News Special Coverage of Space Exploration From 1957 To 2003," 2005

Conversation with Gerry Griffin, 10/28/2009

Conversation with Roy L. Flemming 4/27/2010

Correspondence with James Oberg, 10/29/2015

Correspondence with Susan Dietrich, 8/24/2015

Correspondence with Dave Dalton, 8/26/2015

Correspondences with Jim Behling, 3/17-18/2015, 9/8-9/2015

Correspondence with David Morken, 9/19/2015

Correspondence with Ed Kyle, 1/2/2016

The Saginaw Area GIS Authority

https://archive.org/details/geminiaudiocollection

www.NASAspaceflight.com

www.CollectSPACE.com

Space Hipsters Facebook page

http://www.lib.noaa.gov/collections/imgdocmaps/daily_weather_maps.html

Aaron Mintz, News video archival site (now closed).

Author W. Wes Oleszewski was born and raised in mid-Michigan and spent most of his life with an eye turned toward the space, flight and spaceflight. Since 1990 authored 17 books on the subject of Great Lakes maritime history and lighthouses. Now he has turned his attention toward spaceflight.

Noted for his meticulous research, Oleszewski has a knack for weeding out the greatest of details from the most obscure events and then weaving those facts into the historical narratives which are his stories. His tales of actual events are real enough to thrill any reader while every story is technically correct and highly educational. Oleszewski feels that the only way to teach history in this age of computer and video games is through "narrative." The final product of his efforts are captivating books that can be comfortably read and enjoyed by everyone from the eldest grandmother to the grade-school kid and future pilot or historian.

Born on the east side of Saginaw, Michigan in 1957, Wes Oleszewski attended public school in that city through grade nine, when his family moved to the town of Freeland, Michigan. In 1976 he graduated from Freeland High School and a year later entered the Embry-Riddle Aeronautical University in Daytona, Florida. Working his way through college by way of his own earned income alone, Oleszewski graduated in 1987 with a commercial pilot's certificate, "multi-engine and instrument airplane" ratings as well as a

B.S. Degree in Aeronautical Science. He has pursued a career as a professional pilot as well as one as an author. He holds an A.T.P. certificate and to date has filled more than three logbooks with flight time most of which is in airline category and jet aircraft. Recently he gave up the life of a professional aviator and now enjoys his job as a professional writer.

AUTHOR ACKNOWLEDGEMENTS

This is normally the section where I deliver thanks to libraries, researchers and private collectors, etc. Most of those who need acknowledgement will actually be found in the text, or bibliography, so here I am going to place thanks upon those who are not in the text. There are a few ultra- important folks without whom I would never have produced this series of books. When I first decided to stick my toe into the ebook world I did some forum reading and saw how many newbie e-authors were bashing their brains over one critical area: formatting. So I contacted my long-time friend, fellow professional pilot and author as well as ERAU alumnus, Mark Berry. He had already done some e-books and when asked he put me in contact with Kristina Blank Makansi, founder of Blank Slate Communications which provides Author Services. With her guidance I discovered just how sheltered my life has been while working in print and how much my print publisher has done that I took for granted. She taught me about things such as proper photo DPI, bar codes, ISBNs and countless other areas upon which I would have stumbled. Joining Kristina in the effort was her daughter Elena who took on the task of formatting the interior and placing photos in the book. Together they form a professional team that was instrumental in bringing this book and those that will follow, into reality. I thank them.

I would also like to thank my close friend Brian Nicklas- who, by the way, was also the best man in my wedding back in 1988. Brian is currently with the National Air and Space Museum and aided in the creation of this text by way of technical advice. When one is seeking technical advice, one seeks out the best- so I sought Brian.

WES OLESZEWSKI

Made in the USA
San Bernardino, CA
13 October 2016